by
James R. Payne and Charles R. Phillips

PETROLEUM SPILLS

IN THE MARINE ENVIRONMENT

The Chemistry and Formation of Water-In-Oil Emulsions and Tar Balls

CRC Press
Taylor & Francis Group
Boca Raton London New York

CRC Press is an imprint of the
Taylor & Francis Group, an informa business

First published 1985 by CRC Press
Taylor & Francis Group
6000 Broken Sound Parkway NW, Suite 300
Boca Raton, FL 33487-2742

Reissued 2018 by CRC Press

Library of Congress Cataloging in Publication Data
Main entry under title:

Payne, James R.
Petroleum spills in the marine environment.

Bibliography: p.
Includes Index.
1. Oil Spills. 2. Emulsions. 3. Petroleum–
Biodegradation. I. Phillips, Charles R.
(Charles Robert), 1951- II. Title.
GC1085.P33 1985 628.1'6833 85-13065
ISBN 0-87371-058-4

A Library of Congress record exists under LC control number: 85013065

Publisher's Note
The publisher has gone to great lengths to ensure the quality of this reprint but points out that some imperfections in the original copies may be apparent.

Disclaimer
The publisher has made every effort to trace copyright holders and welcomes correspondence from those they have been unable to contact.

ISBN 13: 978-1-315-89638-0 (hbk)
ISBN 13: 978-1-351-07548-0 (ebk)

Visit the Taylor & Francis Web site at http://www.taylorandfrancis.com and the
CRC Press Web site at http://www.crcpress.com

PREFACE

This book presents an overview of the current understanding of water-in-oil emulsion behavior. It is based on a review originally prepared as a contribution to the NAS Petroleum in the Marine Environment/Update Workshop held in November, 1981. It may or may not reflect the concensus of the workshop participants or the National Academy of Sciences. A summary publication of those proceedings was published in 1985; however, there are currently no plans to publish the background papers per se. Reviews of recently published studies, sitings, and investigations at spills-of-opportunity, as well as results of recent arctic and subarctic oil weathering experiments and observations on the behavior of crude oil in the presence of ice have been added to the manuscript to provide coverage of important research completed between 1981 and May, 1985. This new book contains a tremendous amount of information of interest to petroleum companies, scientists, engineers, environmentalists, chemical and biological oceanographers, and oil spill response coordinators, and this book reflects the authors' desire to make its contents available to a wider audience.

When crude oils or refined petroleum products are released at sea, they are subjected immediately to a wide variety of weathering processes. These processes cause significant changes in the rheological properties of the oil, several orders of magnitude increases in viscosity, factor of two changes in interfacial surface tensions, and appreciable (5 – 10%) changes in oil density. All of these alterations have important implications for the subsequent spill cleanup or containment strategy. The increase in volume due to incorporation of water in emulsions can cause further problems for storing the collected product.

Specific laboratory studies on water-in-oil emulsions with a variety of crude and refined petroleum products are discussed in detail in Chapter 2. This information, along with data on the component concentrations of wax, asphaltene, and other surface-active materials, facilitates reliable predictions on whether or not a particular crude or refined product will form a stable emulsion. Chapter 3 covers case histories of real spill events with discussion of specific studies of water-in-oil emulsions and the chemistry of emulsion formation and behavior. The chemical composition of tar balls, their sources, fate, and global distribution are discussed in Chapter 4. Chapter 5 is an overview of recent attempts to model water-in-oil emulsification behavior. Chapter 6 presents a summary and critical citation review for easy reference to additional details on particular topics presented in the other chapters.

We are grateful to a number of colleagues, friends, and peers who have encouraged us since the NAS review to revise and publish this material in its present form. The original draft of the manuscript was reviewed by Professor James N. Butler of Harvard University and Mr. Gerard P. Canevari of Exxon Research and Engineering Company. Their comments and suggestions are greatly appreciated and have been incorporated throughout the text where appropriate. We must assume responsibility, however, for any omissions introduced in reviewing the rather extensive amount of literature published on the subject.

We are deeply indebted to G. Daniel McNabb, Jr., James Lambach, Robert Redding, and William Paplawsky for analyses of samples from Kasitsna Bay, Alaska, discussed in the book. We would also like to thank Cheryl Fish, Suzanne Goldman, Mabel O'Byrne, Randee Luedecke, and Nancy Burnett for their able assistance in the preparation of this manuscript, and its tables and figures. James Lambach and Randolph E. Jordan were also instrumental in obtaining much of the literature cited and in compiling the various references for the bibliography. Jan Vorhees is thanked for her patience and persistence in waiting for the original manuscript for the NAS review and for her encouragement in prompting JRP to update and publish the current version of this text.

Oil weathering research programs related to the subject of this review have been funded (wholly or in part) by NOAA/Office of Marine Pollution Assessment, Contract No. NA80RAC 00013, Cooperative Research in Investigating the IXTOC I Blowout; NOAA/Outer Continental Shelf Environmental Assessment Program, Contract No. 03-7-002-35213, Preparation of a Literature Review on the Fate and Weathering of Petroleum Spills in the Marine Environment; by the Bureau of Land Management through interagency agreement with the National Oceanic and Atmospheric Administration, as part of the Outer Continental Shelf Environmental Assessment Program, Contract No. NA80RAC00018, Multivariate Analysis of Oil Weathering in the Marine Environment—Sub Arctic; and by NOAA/Outer Continental Shelf Environmental Assessment Program, Contract No. 813-ABC-00062, Development of a Predictive Model for the Weathering of Oil in the Presence of Sea Ice.

James R. Payne, Ph.D., and Charles R. Phillips
April 8, 1985
Division of Applied Environmental Sciences
Science Applications International Corporation
La Jolla, California

James R. Payne is a Senior Chemist/Senior Project Manager in the Division of Applied Environmental Sciences, Science Applications International Corporation, La Jolla, California. He received his Ph.D. in Chemistry from the University of Wisconsin – Madison, and a B.A. with honors in Chemistry from California State University – Fullerton. He was a National Institutes of Health Predoctoral Fellow at Wisconsin and focused his research interests on apoenzyme/coenzyme interactions and the synthesis of coenzyme analogs, including ^{13}C-enriched vitamin B-6, for ^{13}C nuclear magnetic resonance studies of enzyme/coenzyme active site complexes. After graduate school, he received a Woods Hole Oceanographic Institution Post-Doctoral Scholarship, where he undertook research on marine humic acids, incorporation of petroleum hydrocarbons into marine shellfish, and the persistence and metabolism of PCBs in the water column of the North Atlantic.

Later research has centered on development of laboratory methods and large-volume seawater sampling systems for detection of trace-level organics, particularly petroleum hydrocarbons, in seawater and tissues of selected marine species. He was an invited scientist on the NOAA ship RESEARCHER, investigating the IXTOC I oilwell blowout in the Bay of Campeche, and Principal Investigator of a three-year laboratory and field program designed to provide a multivariate analysis and computer model of petroleum weathering in the marine environment. More recently, Dr. Payne's experimental and modeling efforts have been expanded to include the interactions of spilled petroleum with sea ice under arctic conditions. Other research interests include development of chemical and bacterial countermeasures for *in situ* cleanup of contaminated soils, evaluations of oil dispersant effectiveness in low temperature/low salinity environments, and GC/MS analyses of trace level pollutants encountered in methane gas recovery from capped landfills.

Charles R. Phillips is a Staff Scientist with the Division of Applied Environmental Sciences, Science Applications International Corporation in La Jolla, California. He received a B.A. in biological sciences from the University of California, Santa Barbara and an M.A. in marine sciences from San Francisco State University. Graduate research at Moss Landing Marine Laboratory focused on measuring and characterizing the vertical flux of particulate organic material in the coastal marine environment.

Recent research includes participation in Dr. Payne's oil weathering in the marine environment, oil-in-ice and oil-SPM experimental and modeling programs, dispersant effectiveness studies, and environmental monitoring and impact assessment programs. Mr. Phillips has co-authored several recent papers on the environmental weathering of oil, an oil pollution index, and the fate of hydrocarbons from OCS oil and gas development programs in the marine environment.

LIST OF FIGURES

Figure

1 Rheological Properties Data on the Prudhoe Bay Crude Oil Weathering in the Wave Tank Systems 17

2 Flame Ionization Detector—Temperature Programmed Gas Chromatographic Analysis of Tenax-Trapped Volatile Compounds Lost From Fresh Prudhoe Bay Crude Oil After 1 Hour of Weathering (With Turbulence) on Seawater 26

3 Flame Ionization Detector—Temperature Programmed Gas Chromatographic Analysis of Tenax-Trapped Volatile Compounds Lost From Fresh Prudhoe Bay Mousse (80% Water) After 1.5 Hours of Weathering (With Turbulence) on Seawater 27

4 Tenax Trap FID-GC Data on Sub-Arctic Volatile Component Loss From Prudhoe Bay Crude Oil and Mousse on Flow-Through Seawater Enclosures in Kasitsna Bay, Alaska 28

5 Computer Generated Plots of Capillary FID-GC Data on Intermediate Molecular Weight Components Remaining in Prudhoe Bay Crude Oil and Mousse Weathering Under Sub-Arctic Conditions on Flow-Through Seawater Enclosures at Kasitsna Bay, Alaska 29

6 Computer Generated Plots of Capillary FID-GC Data on Higher Molecular Weight Components Remaining in Prudhoe Bay Crude Oil and Mousse Weathering Under Sub-Arctic Conditions on Flow-Through Seawater Enclosures at Kasitsna Bay Alaska 31

7 Location of the IXTOC-I Blowout and Direction of Oil Slick Transport in the Gulf of Mexico 58

8 IXTOC-I Campeche Oil Spill Cruise; Expanded Wellhead Region Showing Location of Turbid Water Boundary 62

9 Relative Abundance of Alkyl-Substituted Polynuclear Aromatic Hydrocarbons in IXTOC Crude Oil Collected ½ Mile from the Wellhead 71

10 Abundance of N-Alkanes (Relative to nC-20) in Beached Mousse from Laguna Madre (RIX-23), Mousse Flakes Collected 16–18 Miles from the Wellhead (P-13) during the *GW Pierce* Down-Plume Transect, and IXTOC Crude Collected ½ Mile from the Wellhead 72

11 Mousse/Oil Relative Viscosity Ratio of Six Test Crude Oils as a Function of Water Content 101

12 Computer Model—Predicted and Observed Time-Dependent Uptake of Water into Ekofisk Oil as a Function of Wind Speed and Sea-State 103

LIST OF TABLES

Table

1 Mousse Formation Experiments using a Variety of Fresh and Artificially Weathered (Topped) Crude Oils in Laboratory, Outdoor Test Tank, and Field Experimental Spills 6

2 Other Oils Which Have Demonstrated Water-in-Oil Emulsion Tendency 9

3 Time-Series Water Column Concentrations of Dissolved and Dispersed Hydrocarbons from Fresh Prudhoe Bay Crude Oil and Mousse Weathering on Flow-Through Seawater Enclosures (Turbulent Regime) at Kasitsna Bay, Alaska 34

4 Inhibition of Stable Mousse Formation by the Addition of Chemical Dispersants 36

5 Chemical Studies of Oil/Mousse Behavior in Major Oceanic Spills and Blowouts 42

6 Selected Component Ratios for the Dissected Mousse Sample Beached at Laguna Madre (Station RIX 23) 73

7 Summary of Tar Ball Distributions and Concentrations on the World Oceans 84

8 Summary of Stranded Tar Ball Distributions and Concentrations on Beach Surfaces 95

TABLE OF CONTENTS

Chapter

1. **INTRODUCTION** 1

2. **LABORATORY STUDIES OF FORMATION AND STABILITY OF WATER-IN-OIL EMULSIONS** 3

   Background 3
   Specific Studies 5
   Mousse Formation in the Presence of Ice 18
   Identification of Emulsifying Agents Responsible for Mousse Formation 21
   Physical and Behavioral Properties of Water-in-Oil Emulsions 25
      Evaporation 25
      Combustibility 30
   Breaking and Interaction of Laboratory Mousse with Dispersants 33
   Bacterial Utilization of Laboratory Generated Mousse 40

3. **SELECTED CASE HISTORIES OF THE MORE DETAILED CHEMISTRY STUDIES OF MOUSSE BEHAVIOR AND LONG TERM FATE IN NEAR-COASTAL AND OPEN OCEAN OIL SPILLS/BLOWOUTS** 41

   Torrey Canyon 41
   Tanker Arrow 43
   Metula 46
   Ekofisk Bravo Blowout 47
   US/NS Potomac in Melville Bay, Greenland 50
   Amoco Cadiz 51
   IXTOC I Blowout, Bay of Campeche, Gulf of Mexico 57
      Background 57
      Observations at the Wellhead 61
      Subsurface Transport and Weathering of IXTOC Oil 66
      Personal Observations of Micro-Scale Mousse Agglomeration 67
      Fate of Stranded IXTOC Mousse Along the Southeast Texas Coastline 74
   Burmah Agate 76
   Alvenus 77
   Open Ocean Field Tests of Spilled Petroleums 78
   Conclusions 81

4. **TAR BALL FORMATION AND DISTRIBUTION** 83

Global Distribution of Pelagic Tar 83
Chemical Composition of Tar Balls 86
Sources of Pelagic Tar 90
Fate of Pelagic Tar at Sea 91
Fate of Beached or Stranded Tar Balls 94

5. **ALGORITHMS AND COMPUTER PROGRAMS TO SIMULATE THE FORMATION OF WATER-IN-OIL EMULSIONS** 99

6. **SUMMARY, CONCLUSIONS, AND CRITICAL CITATION REVIEW** 113

Laboratory Studies 113
Physical Properties of Water-in-Oil Emulsions 115
Treatment of Mousse with Dispersants 115
Case Histories of Real Spill Events 116
Tar Ball Distributions and Chemistry 118
Mathematical and Computer Modeling of Mousse Behavior 119

BIBLIOGRAPHY 121

INDEX 141

CHAPTER 1

INTRODUCTION

A better understanding of the phenomenon of water-in-oil emulsification and tar ball formation from petroleum spills at sea is critical to our ability to predict, control, and mitigate the environmental impacts of petroleum hydrocarbon spills in marine and coastal waters. Stable water-in-oil emulsions or "mousse" complicate clean-up strategies and logistics because the more viscous emulsions present formidable problems in skimming, pumping, and recovery operations. Emulsified oils also require an inordinate amount of space in transport and intermediate storage due to the increase in volume from water incorporation. Additionally, in the final stages of disposal, certain water-in-oil emulsions may resist more convenient and conventional disposal mechanisms such as burning.

When crude oil and many refined products are released at sea they are subjected immediately to a series of weathering processes including: spreading; evaporation and dissolution of selected lower molecular weight components; dispersion of whole oil droplets into the water column; coalescence and return to the surface slick of those droplets with entrapment of seawater; photo-, microbial- and auto-oxidation; and emulsification and tarball formation. The rates of these concomitant processes are inextricably linked (and in some cases compete) with one another. They are also dependent on the type and amount of oil spilled (component concentrations), environmental conditions (water and air temperature, wind speed, and turbulence regime -- sea state), and man's own efforts to control or disperse the slick through the application of dispersants/demulsifiers and/or sinking agents.

Most crude oils and refined petroleum products have specific gravities less than one and will not readily sink after initial release. However, the combined effects of natural and enhanced weathering processes alter the density, viscosity, pour point, and volume of these products such that ultimate cleanup and containment strategies must be capable of handling an incredible variety of situations. During formation of water-in-oil emulsions, products of higher density and viscosity which contain up to 70 to 80% water (dispersed as sub-micrometer to 50 micrometer droplets in the continuous oil phase) can be generated and, as such, burning may become more difficult due to the high water content and chemical dispersion may prove to be impossible. Because of their greater density, however, water-in-oil emulsions, and tar balls generated from such mousse, may be more susceptible to submersion which ultimately enhances dispersion of oil slicks.

Numerous investigators have suggested that mousse formation and stability are influenced by the presence of surfactant materials such as asphaltenes, waxes, organometallics, and nitrogen, sulfur and oxygen (NSO) compounds which are important in preventing water-water droplet coalescence within the emulsion. It has been found that the more viscous oils tend to form more stable emulsions and water-in-oil emulsions form more rapidly under lower temperature (higher viscosity) than higher temperature conditions. Turbulence has also been demonstrated to be critical in mousse formation. At this time, however, no single explanation accounts completely for all of the observations, and not all of the mechanisms of mousse formation and stability are understood.

In this review, attempts have been made to provide a broad view of the subject. Topics include discussions of laboratory and field test-tank (wave and mixing chamber) experiments used to examine specific factors associated with mousse formation. The importance of oil composition and different turbulence regimes, as well as discussions of studies of mousse formation and behavior in real spill situations are included. Several major spill incidents are considered with regard to observed and documented mousse and tarball formation, stability, and fate. Whenever possible, correlations are made between real spill situations and laboratory simulations. The occurrence, distribution, and chemistry of tarballs from other sources are briefly considered; however, not as much emphasis has been placed on this subject due to the highly variable levels of tarballs in the world's oceans and their somewhat limited long-term environmental impact.

Finally, a brief review presents recent attempts to simulate mousse formation and behavior through mathematical and computer modeling. These models generally are coupled to, or based upon, laboratory wave tank and mixing chamber experiments, although several attempts to model field observations with computer predictions (hind casting) have been completed.

CHAPTER 2

LABORATORY STUDIES OF FORMATION AND STABILITY OF WATER-IN-OIL EMULSIONS

BACKGROUND

Before undertaking a discussion of water-in-oil emulsion formation and stability, it is necessary first to review several general aspects of emulsions and emulsification. A more comprehensive treatment of the subject is presented by Twardus (1980). In general, an emulsion is defined as two immiscible liquids wherein droplets of one phase (the dispersed or internal phase) are encapsulated within sheets of another phase (the continuous or external phase). When crude oil or petroleum products are spilled at sea, two basic forms of emulsions are possible. The first is an oil-in-water (O/W) emulsion in which oil droplets are dispersed and encapsulated within the water column. The second is a water-in-oil (W/O) emulsion in which droplets of water are dispersed and encapsulated within the oil. This second mixture is generally referred to as mousse in the literature. For either type of stable emulsion to form between two liquids, three basic conditions must be met: (1) the two liquids must be immiscible or mutally insoluble in each other; (2) sufficient agitation must be applied to disperse one liquid into the other; and (3) an emulsifying agent or combination of emulsifiers must be present.

During emulsification, the interfacial area between two liquids increases. Liquids tend to minimize this surface area, therefore, an emulsifying agent and work (or energy) are required for emulsification to proceed. In theory, the amount of energy required to increase the surface area can be calculated if the interfacial tension between the two liquids is known (Becher, 1955). In open ocean and coastal oil spills, sufficient energy to satisfy this requirement typically is provided by wind, waves, and currents. Nevertheless, stable water-in-oil emulsions also have been observed to form with certain oils even on very calm seas. The emulsifying agent may be any surface active substance which can form a thin interfacial film between the two liquids and maintain the emulsion by minimizing the contact, coalescence, and aggregation of the internal dispersed phase. For emulsions to form in the absence of external agitation, the interfacial tension between the two liquids should be reduced to approximately 0.5 dynes/cm, whereas only approximately 5 dynes/cm are needed for emulsions formed with agitation. The surfactant should surround the dispersed droplet as a non-adhering film and should have a molecular structure in which the polar end is attracted to the water and the non-polar end is attracted to the oil. Surfactants should be relatively more soluble in the external phase so that they are readily available for adsorption around the internal phase. These surfactants also may impart an electro-kinetic potential and influence the viscosity of the emulsion formed. Finally, the surfactant material must stabilize the emulsification process while present in relatively small quantities.

Depending on the chemical composition of the surfactant, emulsion stability can either increase or decrease. For example, materials containing mono-valent ions have been shown to stabilize oil-in-water emulsions, whereas surfactants containing poly-valent ions can stabilize water-in-oil emulsions. A number of materials are present in crude oils which stabilize water-in-oil mixtures (these will be discussed later in greater detail). In general, however, unrefined oils have relatively higher portions of water-in-oil emulsifying agents than oil-in-water emulsifying agents. Thus, while both types of emulsions can form in petroleum spills, the majority of the emulsion would be the water-in-oil type. Furthermore, oil-in-water emulsions are inherently unstable, and they have been shown to revert to water-in-oil mixtures. Effects of shear rate, temperature, and oil concentration on the formation of oil-in-water emulsions were studied by Mao and Marsden (1977) using California crude. They noted that increases in temperature and/or oil concentration enhanced the conversion of oil-in-water emulsions to water-in-oil emulsions.

In water-in-oil emulsions, asphaltene substances, porphyrin complexes, and waxes act as natural emulsifying agents stabilizing W/O mixtures (Berridge et al., 1968a, 1968b; Cairns et al., 1974; Canevari, 1969; Frankenfeld, 1973). Presumably these agents provide the required film around the water droplets which resists rupture, thus preventing water-water coalescence (Canevari, 1982). The size distribution of water droplets in W/O emulsions, discussed in greater detail later, is also important.

The stability of water-in-oil emulsions is dependent on a variety of factors, including: the presence or absence of the emulsifying agent, viscosity (influenced greatly by temperature), specific gravity, water content, and the age of the emulsion. Essentially, the stability of a W/O emulsion could be defined as the resistance by the dispersed water droplets against coalescence. This definition is based upon the phenomenon of Brownian movement, such that the emulsions having a high specific gravity and viscosity would tend to be more stable because movement of the water droplets theoretically would be reduced. As noted above, increases in temperature which result in reductions of viscosity or increases in the water droplet concentration in the continuous petroleum phase would increase the probability of collision and coalescence, thus destabilizing water-in-oil emulsions.

SPECIFIC STUDIES

A number of laboratory experiments have been undertaken in mixing chambers and wave tanks to study the formation and behavior of mousse. Evaporation and dissolution typically were allowed to occur to simulate ambient environmental conditions. In almost all instances, hydrocarbons with molecular weights less than nC-11 to nC-12 (distillation range 200° to 225°C) were lost during the initial stages of weathering, as observed in studies of open ocean and near-coastal spills. The results of these studies and the physical properties and selected chemical characteristics of the crude oils and resultant water-in-oil emulsions are summarized in Table 1. Table 2 lists several additional oils which have demonstrated water-in-oil emulsification tendencies, but for which only limited data are available.

Berridge el al. (1968a, b) studied the effects of chemical composition of the starting crude and evaluated mousse formation potential and stability for seven crudes that were selected to give a representative sampling of oils likely to cause marine pollution. Specific gravities of the crudes ranged from 0.829 for Libyan (Brega) crude to 0.896 for the Venezuelan (Tia Juana medium) crude. Sulfur contents for the crudes range from 0.2 to 2.5%, and kinematic viscosities at 100°F ranged from 4.13 to 25 centistokes. Pour points for the selected oils ranged from -34° to +7°C, and wax contents were found to vary inversely with the specific gravity, ranging from a high of 11% by weight for the Libyan crude to 4.8% by weight for the Tia Juana crude. Additional characterization data are presented in Table 1. Asphaltenes were found to increase in weight percent from 0.13 to 3.5 and were roughly inversely proportional to the wax content. Vanadium content increased with the increase in specific gravity and percent asphaltenes. Residues with components having boiling points greater than 370°C ranged on a weight percent basis from 35 to 57% for the crudes studied, and interestingly, the residue pour points decreased from 100° to 50°C from the light to the heavier crudes. Thus, when spilled at sea, crudes such as the Libyan Zelten (Brega) and the Nigerian light, which have fairly low percentage ranges of residues greater than 700°, will be removed relatively rapidly by evaporation. Evaporation is particularly effective for weathering Zelten crude which contains 31% by weight of components which boil below 200°C. Heavier crudes, such as the Tia Juana medium (78% residue boiling greater than 370°C), contain only a small fraction which distills at low temperature; thus, they would evaporate very slowly and would not be expected to weather appreciably by evaporative processes.

The weathered residues obtained from the evaporative processes acting on all the oils studied had higher specific

Table 1. Mousse formation experiments using a variety of fresh and artificially weathered (topped) crude oils in laboratory, outdoor test tank, and field experimental spills.

	Initial Oil Properties								Water-In-Oil Emulsion (Mousse) Properties					
Product Tested	Specific Gravity	Viscosity	Pour Point °C	Wax % Weight	% Asphal-tenes	% Sulfur	V (ppm)	Ni (ppm)	Stable Mousse Formed/ Appearance	Final % Water	Viscosity	Pour Point °C	Bacterial Growth Noted (6 weeks)	References
Libyan (Brega)	0.829	4.13*	7.2	11.4	0.13	0.21	5		Borderline/ dark brown-waxy	78.3			Fairly heavy	Berridge et al., 1968 a,b
Nigerian Light	0.867	5.16*	-15.0	8.5	0.05	0.19	5		Borderline/yellow brown-granular	77.3			Heavy infestation	Berridge et al., 1968 a,b
Iranian Light (Agah Jari)	0.854	5.6*	-20.6	7.0	0.7	1.33	36		Rigid/foamy	79.1			Fairly heavy	Berridge et al., 1968 a,b
Iranian Heavy (Gach Saran)	0.869	8.83*	-12.2	6.7	1.9	1.58	107	37	Rigid/mid-brown	77.3			Very few	Berridge et al., 1968 a,b
Iraq (Kirkuk)	0.845	4.75*	-34.4	6.5	1.3	1.88	25		Rigid/dark brown	78.3			Very few	Berridge et al., 1968 a,b
Kuwait	0.869	9.6*	-31.7	5.5	1.4	2.5	27	9	Rigid/mid-brown	79.1			Some present	Berridge et al., 1868 a,b
Venezuelan (Tia Juana Medium)	0.869	25.0*	-34.4	4.8	3.05	1.54`	170	16	Rigid/dark oily brown	73.8			Very few	Berridge et al., 1968 a,b
Bunker C	0.990	2.8×10^7 cP @ 10°C	+7						Rigid and sticky	67	2.9×10^7 cP			Berridge et al., 1968 a,b Twardus, 1980
Light Arabian Crude	0.99	6.04 cS @38 °C									500-800 cS @ 20°C			Solsburg, 1976 Twardus, 1980
Norman Wells in	0.83 0.93*	8.68 cP @10°C	-85 -8**						Unstable	50	110 cP @10°C 240**	-48		Twardus, 1980 (additional data original reference).
Sweet Blend	0.83 0.94**	14.2 cP @10°C	-35 -12**						Unstable	60	450 cP @10°C 520**	-48		Twardus, 1980
Sour Blend	0.83 0.94**		-50 +18**								3000**			Twardus, 1980
Bow River	0.90 0.99**		-27 0**								2150**			Twardus, 1980

Table 1. Mousse formation experiments using a variety of fresh and artificially weathered (topped) crude oils in laboratory, outdoor test tank, and field experimental spills. (Continued)

Product Tested	Initial Oil Properties								Water-In-Oil Emulsion (Mousse) Properties					
	Specific Gravity	Viscosity	Pour Point °C	Wax % Weight	% Asphal-tenes	% Sulfur	V (ppm)	Ni (ppm)	Stable Mousse Formed/ Appearance	Final % Water	Viscosity	Pour Point °C	Bacterial Growth Noted (6 weeks)	References
Lloydminster	0.90	86.8 cP @10°C	-52						Stable	60	2800 cP @10°C			Twardus, 1980
	0.98**		-9**								2675**			
Weyburn Midale	0.89	29 cP @10°C	-28						Stable	60	4150 cP @10°C	-2.5		Twardus, 1980
	0.99**		-3**								1250**			Twardus, 1980
Topped 200+ Kuwait (dewaxed & deasphaltenized)				Added % wax & % asphaltenes										Bridie et al. 1980 a,b
				0	0				Unstable emulsions***					
				5	0									
				0	1.4				Unstable emulsion****					
				5	1.4				Stable mousse					
				5	0.14				Stable mousse					
Lube Oil									No emulsion	<1				Bridie et al., 1980 a,b
Lube Oil + 5% asphaltenes/wax mix									Stable emulsion	54	0.01-0.03 um water droplet size			Bridie et al., 1980 a,b
Lube Oil +10% asphaltenes/wax mix									Stable emulsion	67	0.01 um water droplet size			Bridie et al., 1980 a,b
Venezuelan (Guanipa)	0.859			-	5.2	1.66	105	18	Mousse and tarballs and flakes formed after few days in wave tank at 20°C (85% water-in-oil in first few days; dropped to 75% over 4 months). Specific gravity approached 1.0.					MacGregor and McLean, 1977
Libyan (Saria)	0.843			20	0.15	0.14	0.5	5						MacGregor and McLean, 1977
Algerian (Zaraitine)	0.818			5.3	0.08	0.09	1	1	Unstable emulsion					MacGregor and McLean, 1977
Nigerian Medium	0.892			4.6	0.1	0.25	0.8	7						MacGregor and McLean, 1977

Table 1. Mousse formation experiments using a variety of fresh and artificially weathered (topped) crude oils in laboratory, outdoor test tank, and field experimental spills. (Continued)

	Initial Oil Properties								Water-In-Oil Emulsion (Mousse) Properties					
Product Tested	Specific Gravity	Viscosity	Pour Point °C	Wax % Weight	% Asphal-tenes	% Sulfur	V (ppm)	Ni (ppm)	Stable Mousse Formed/ Appearance	Final % Water	Viscosity	Pour Point °C	Bacterial Growth Noted (6 weeks)	References
Alberta Crude	0.829	8.25 cP @20°C							Stable mousse	70-80% in 30-minute wave tanks	100-200 cP			Mackay et al., 1979
350° Topped Kuwait	0.958	208 cS @50°C	+8		2.0	7% total 3.7	50	15	Stable mousse	50% after 8-10 weeks	2350 cS @50°C	13-15		Davis and Gibbs, 1975
Prudhoe Bay	0.893	19 cS	-10	-	23	-	28.3	13.5	Stable mousse/ light brown	55	2800 cS after 12 days 7800 Cs after 4 months in subarctic wave tank experiments			Payne et al., 1981a; 1983a
									Stable mousse	65	25,000 cS after 11 hrs 30,000 cS after 36 hrs in the presence of ice			Payne et al., 1984b
Gasoline									No emulsion or mousse	Nil				Berridge et al., 1968b
Kerosene									No emulsion or mousse	Nil				Berridge et al., 1968b
Auto Diesel									No emulsion or mousse	Nil				Berridge et al., 1968b
Marine Diesel	0.83 9.87**	10 cP @10°C	-15 -8**						No emulsion or mousse	Nil		17**		Twardus 1980; Berridge et al., 1968b
Lube Oil 600									Unstable emulsion					Berridge et al., 1968b
Lube Oil 2500									Fluid emulsion but no mousse					
Heavy naph Lube									Fluid emulsion hut no mousse					

* Kinematic viscosity (cS) at 100°F
** Specific gravity and pour point after 4 weeks pan evaporation under atmospheric conditions (no water added except for occasional precipitation).
*** 93% of water shed after standing 15 minutes
**** 88% of water shed after standing 15 minutes

TABLE 2

Other Oils Which Have Demonstrated Water-in-Oil Emulsion Tendency
(from Bridie et al., 1980a)

Crude Oil	Source	Mousse Formed	Flow Properties	Spreading on Water at 10° C
Brent	N Sea GB	+	Viscous	-
Ekofisk	N Sea Norway	+	Unstable	-
Auk	N Sea GB	+	Viscous	-
Kuwait	Kuwait	+	Paste	-
Nigerian Medium	Nigeria	+	Low Visc	+
Qatar Marine	Qatar	+	Viscous	-
Cabimas	Venzuela	-	-	+
Iranian heavy	Iran	+	Paste	-

gravities, viscosities, sulfur, metal, and wax content than the original crudes. For example, Kuwait crude contains approximately 27% by weight residual materials with a boiling point above 1,000°C and a specific gravity of 1.028. Similarly, an Iranian heavy crude residue has a specific gravity of 1.027. Therefore, both of these residues have densities greater than seawater (1.025 g/cm^3) and would be expected to sink relatively easily in the marine environment.

Rigid, stable emulsions were formed with most of the oils tested, with the exception of Brega and Nigerian light crude which were classfied as marginal. Emulsion colors ranged from mid-brown to dark oily-brown to a yellow brown granular substance with the Nigerian light crude. Residual fuel oil (Bunker C) also formed a stable mousse. In contrast, no mousse could be generated using distillates such as gasoline, kerosene, and diesel oils. Lubricating fluids did not form stable "mousse", but emulsions that were either unstable or fluid could be generated. In general, variation in the size of water droplets appeared to correlate with stability; the Brega and Nigerian crudes contained the largest water droplets and exhibited the least stability and greatest potential for the water and oil to separate into distinct phases. The more stable emulsions contained water droplets with diameters less than or equal to one micrometer.

The effect of salinity was also investigated by Berridge et al. (1968a, b). Mousse type emulsions were obtained with Kuwait crude and Tia Juana crude with water contents ranging from 74 to 80% regardless of salinity. A stable emulsion formed in all cases, but the appearance of the mousse ranged from a mid-brown for Kuwait crude with seawater to a mid-gray-brown using tap or distilled water. Tia Juana crude with seawater formed a very dark brown mousse, whereas the mousse formed with tap water and distilled water was nearly black. To evaluate other factors affecting mousse formation, oil and water were filtered through Whatman #1 filter paper to remove particulates above 100 microns, and in several instances the water was sterilized by the addition of 500 ppm dichlorophen. In general, these procedures did not affect the water content or mousse appearance.

Mousse stability was measured by placing 425 ml of mousse from the different crudes on glass plates and allowing them to weather naturally under environmental conditions. Identical samples were also placed in 2 gallon buckets and agitated from below with bubbles. Stable mousse on the glass produced an oil fringe while maintaining an overall rigidity, whereas less stable mousses deformed or slumped and flowed off the glass under the influence of wind, gravity, and rain. Estimated losses due to weathering of so-called beached mousse ranged from 10% for Kuwait crude to 80% for the Brega crude, which formed a much softer, waxy mousse. Nigerian crude oil mousse was very granular on the plate and formed a waxy or oily and foamy mousse. Microbial populations were observed to grow with Kuwait mousse, whereas very few bacteria were present for Tia Juana, Iraq, Kirkuk, and Gach Seran crudes. Heavy bacterial infestation was observed with the Nigerian light, the Brega crude, and the Iranian (Aja Jari) crudes. Bacterial infestation on samples of mousse floating on aerated seawater was observed, but the most stable mousses again appeared to have the lowest bacterial growth. During experimental periods ranging up to six weeks there was little evidence to suggest that bacterial growth was sufficient to remove the mousse. In fact, there was some evidence with the mousse "stranded" on glass plates suggesting that the presence of bacteria marginally increased stability, although the presence of particulate material had no effect. With regard to water loss on stranded mousse, the two lighter crudes (Brega and Nigerian) exhibited the greatest percent water loss and also had the highest infestation of bacteria. The lack of bacterial activity, noted with several of the mousse samples on the aerated seawater, was attributed to the lack of phosphate and other nutrients or possibly to toxic materials contained in several of the crudes.

Of all the parameters studied, specific gravity and kinematic viscosity were most strongly correlated with mousse stability. Interestingly, oil pour point was not correlated with mousse stability. With regard to chemical composition, the percent residue boiling above 370°C, the asphaltene content, and the vanadium content all showed a definite correlation, whereas acidity, sulfur content, and wax content showed little or no significant correlation with mousse stability.

Davis and Gibbs (1975) used a 350°C topped Kuwait crude (to eliminate viscosity and density changes due to evaporative losses) to study long-term oil and mousse behavior over a two year period in closed and flow-through exposure tanks in Portsmouth, England. The tanks were filled with seawater and 23 liters of oil were added. The authors stated that in retrospect it would have been better to add less oil and leave some of the water surface uncovered because the oil proved to be a barrier to oxygen transfer, leading to oxygen depletion in the closed tanks. Initial oil thickness was about 0.7 cm, but after taking up water the "mousse" reached an ultimate thickness of about 1.4 cm. The most drastic changes occurred during the first nine weeks when water content in the mousse approaching 50%. Vanadium and nickel were not lost over the two year period of the experiment; however, oxygen content increased in the oil from 0.2% in the crude to 3.9% and 2.8% in mousse in the flow-through and closed tank systems, respectively. The pour point of the oil increased from 8° to 11° during the first day, as the water content increased from 0 to 0.8%. After one month the water content increased to 10%, and the pour point of the oil was 15°C. The final 50% water-in-oil mixture was obtained after two months and remained relatively stable for the two year period. Viscosity increases from 216 to 2,350 centistokes at 50°C, the specific gravity of the residue changed from 0.9525 to 0.9825, and the percent asphaltenes increased by a factor of four from 2 to 8%. Polar constituents also increased from 7 to 16%, presumably due to incorporation of oxygenated products. N-alkane components decreased rapidly in the open tank due to microbial degradation. Time series graphs of water content, pour point, percent asphaltenes, specific gravity, and viscosity were presented for the water-in-oil emulsion over the two year period. With regard to mass balance, it was observed that no net loss of the mousse occurred over this time, and the very slow rate of degradation of the water-in-oil mousse over the two year period was attributed to the limited diffusion of oxygen and/or minerals into the mousse. Thus, it was suggested that, aside from removal of n-alkanes in the open system, microbial biodegradation was not a major factor causing observed changes in mousse properties.

MacGregor and McLean (1977) investigated the weathering behavior of Venezuelan (Guanipa) crude oil on synthetic seawater in a fiberglass tank equipped with a wave generator and a controlled radiation system located in an environmental chamber held at 2°C. The crude oil studied in this experiment had negligible wax content, but asphaltenes were measured at 5.2%, and elevated trace metal concentrations (vanadium-105 ppm and nickel-18 ppm) were measured in the oil. The total sulfur content in the original oil was 1.6%, and the specific gravity was 0.859 grams/ml. Evaporation removed the largest quantity of material, and the rate of evaporation was observed to vary directly with exposure time to solar radiation. Losses due to evaporation ceased after 400 hours with approximately 85% of the oil remaining. Only minimal amounts of oil were lost by sinking or solution, although the relative magnitude of these losses were observed to increase with time. The most notable change was a rapid formation of stable water-in-oil emulsions which formed discrete lumps or tar balls within a few days after the spill. During the four month period of the experiment, these tar balls remained very stable and weathering effects were drastically curtailed because oil contact at the air-sea interface was reduced. During mousse and tar ball formation a rapid increase in water content to 85% occurred within the first few days, but then decreased to approximately 75% at the end of the experiment. The specific gravity was noted to increase rapidly, leveling out at approximately 1.0 relative to the seawater density of 1.027 at 2°C. When water content was taken into consideration, the actual increase in crude oil specific gravity (due to evaporation and dissolution alone) was from 0.85 to approximately 0.93. After about two weeks the tar balls had a flake-like appearance and were similar to tar balls observed in the Potomac and IXTOC spills (which will be discussed later under Case Histories). Water content was noted to vary with tar ball size, with the smaller lumps having lower concentrations of water than the larger ones. This difference was attributed to entrapment of pockets of water, in the form of droplets, inside the larger tar balls.

Concentrations of oil dispersed in the water or in true solution increased rapidly to approximately 2 ppm, and then continued to increase with time up to 500 hours when the oil concentration in the water reached 15 ppm. Only 2 to 3 ppm was expected to be in true solution, with the dispersed oil fraction comprising the difference. The authors analyzed both the outer surface and the inner materials of the tar balls, and found no significant difference in the n-paraffin distributions. Nickel concentrations were found to decrease approximately 30% at the end of the experiment, whereas vanadium concentrations increased by about 10 to 15%. Because of the lack of differences in external and internal alkane composition in the tar balls, the authors attributed tar ball formation to mechanical break-up of the parent mousse from available wave energy rather than from additional or extended chemical weathering effects.

Nagata and Kondo (1977) studied the artificial weathering of five crude oils in 6m by 2m by 1m deep tanks where wind, waves, and rain were allowed to interact with the oil. Changes with time in physical and chemical properties were measured for Arabian light crude oil, Iranian heavy crude oil, Kuwait crude oil, an unspecified heavy oil A (50°C viscosity below 20 centistokes and a flash point of 60°C), and heavy oil B (50°C viscosity of 20-50 centistokes and a flash point of 60-70°C). Twenty-one day weathering experiments were conducted and measurements were made of specific gravities, viscosities, and the amount of water incorporated, in addition to gas chromatographic analyses of the oil. The heavy oil A showed the least change; heavy oil B and Arabian light crude oil exhibited similar and intermediate changes, and Iranian heavy and Kuwait crude showed the greatest changes, particularly during the first three days. Specific gravities for these mixtures ranged from 0.87 to 0.98 after 7 days; constant values were obtained after 21 days when an oil-in-water type emulsion was formed. The specific gravity closely paralleled the change in the amount of water in the oil, with the Iranian heavy crude and Kuwait crude showing the highest increases which paralleled increases in specific gravity. It was also stated that evaporation was greatest for non-emulsified oils. Microbial degradation after 5 days reportedly removed approximately 30% of the hydrocarbons below nC-15. Hydrocarbons above nC-15 remained unchanged after five days, but about 50% of the n-alkanes were degraded after 15 days. Photochemical decomposition studies illustrated that secondary and tertiary paraffins were more easily decomposed relative to normal n-alkanes. This was attributed to the lowered activation energies encountered during the oxidation process in breaking tertiary C-H bonds. A number of aromatic compounds were also removed by photo-decomposition; these included anthracene, phenanthrene, 1,2-benzoanthracene, chrysene, fluorene, pyrene, 3,4-benzo-a-pyrene, benzothiophenes and dibenzothiophenes. All of these compounds decomposed, although, decomposition rates of the individual compounds differed appreciably from one another. Anthracene, phenanthrene and 1,2-benzoanthracene decomposed rapidly relative to the other aromatics, and sulfur containing hetero-aromatic compounds showed the same approximate degree of decomposition as compounds with anthracene rings when pure compounds were tested. When a mixture of the hetero-atomic and aromatic compounds was examined, the aromatic coupounds had decomposed quickly, whereas the sulfur containing hetero-aromatic materials remained longer.

Bocard and Gatellier (1981) generated water-in-oil emulsions with Arabian light crude oil topped at 150°C, Safanya, and heavy fuel oil. Generally, viscosities greater than 10,000 centipoise were obtained with the emulsions from all three oils; viscosity ranged from 10,000 to 20,000 cP for the Arabian light, from 20,000 to 50,000 cP for the Safanya, and from 60,000 to 80,000 cP for the heavy fuel oil depending on mixing speed. The viscosities were shown to decrease as the shear rate increased. All of the measurements indicated a plastic-like behavior for the emulsions. Microscopic examinations of the 75% water in Arabian light crude oil mixture showed that, as with other studies of mousse formation, most of the water droplets were from 1-10 micrometers in diameter. These authors found that Nigerian light or Zarzatine (both low asphalt content crudes) did not form stable emulsions. They noted, however, that the higher viscosity oils apparently formed emulsions independently of asphalt content.

Twardus (1980) studied the characteristics of water-in-oil emulsions for eight oils. The stability of the resultant mousses versus water content is presented in Table 1. It is interesting to note that observed increases in viscosity due to mousse formation were very similar to viscosity increases associated with simple evaporation-in-pan experiments conducted with the same crudes and petroleum products (Twardus, 1980). In the evaporation experiments no seawater was added, although the pans did accumulate some water from rain and snow fall. Water-in-oil emulsions formed with several of the crudes and with Bunker C fuel oil after the snow and rain had accumulated. Evaporative loss of hydrocarbons (primarily within the nC-4 to nC-14 range) in the pan evaporation experiments was greatest during aging intervals of six hours, one day, and two days. For longer aging periods (1 to 4 weeks), hydrocarbon evaporation occurred at much lower rates. Evaporation of components from marine diesel fuel oil and from Bunker C was relatively insignificant in the prevailing sub-arctic weather conditions, primarily due to the low percentage of volatile hydrocarbons in these fuel oils.

The percent water incorporated into the emulsions was critical for the stability of mousse formed with Norman Wells crude and Sweet Blend crude. With the first mixture, mousse that exhibited marked stability for several hours could only be formed with 50:50 and 40:60 water-to-oil ratios. For Sweet Blend crude, only 60:40 water-to-oil and 50:50 water-to-oil ratio produced mousse that was stable for several days, but with higher water concentrations the mousse separated following formation. For the other oils tested, Bow River, Lloydminster, and Wabern-Midale, stable mousse was formed at water-to-oil ratios ranging from 20% to 60%. As in other experiments, no stable mousse could be formed using marine diesel fuel. The pour points for the emulsions were generally higher than those of the fresh oils, but for the Norman Wells and Sweet Blend crude oil water-in-oil mixtures, the pour point was less than that of the residue obtained by simple evaporation. For the Wabern-Midale crude oil mousse, the pour point was very similar to that obtained from the pan evaporation experiments in the absence of added seawater.

In order to minimize the problems of density changes due to evaporation, Mackay et al. (1979, 1980) used a 25% weathered Alberta crude oil and measured changes in viscosity as a function of water uptake. Their emulsion studies were conducted in wave tanks, and represent perhaps the most thorough examination of the competitive processes of oil-in-water dispersions versus water-in-oil emulsion behavior. During the first ten minutes after a spill of Alberta Crude oil (SG 0.829, viscosity 8.25 cP) the dominant weathering process was dispersion of oil droplets into the water column, with dispersed oil concentrations reaching 20-50 mg/l. It was speculated that concentrations would increase and approach the limiting values at which dispersion and coalescence rates would be equal; however, this was not observed. Instead, after approximately 20 minutes concentrations in the water columm decreased, and then reached a constant level of 10 mg/l after one hour. This drop in concentration coincided with a significant change in oil appearance, water content, and viscosity. The water content in the oil increased to 70 to 80% volume within 30 minutes, and the viscosity correspondingly increased to 100 to 200 centipoise. Additional details of their studies as they relate to modeling water-in-oil emulsion formation are presented in Chapter 4.

In a study examining the weathering of Prudhoe Bay crude oil under ambient sub-arctic weather regimes, Payne et al. (1983 and 1984a) added oil to 2,800 liter, flow-through, wave tank systems to evaluate changes in chemical and rheological properties with time. Component-specific compound concentrations in the oil and in the water column in the wave tank systems were measured in triplicate, as well as changes in oil density, viscosity, percent water incorporated, and interfacial (oil/air and oil/water) surface tensions. Figure 1, from Payne et al. (1983), summarizes the changes in rheological properties of the oil/mousse observed during the first 12 days of the experiments. Water was not entrained in the oil for the first twelve hours of the spill, although appreciable dispersion of oil droplets into the water column was noted. After approximately twelve hours water content in the oil increased at a constant rate, and reached a maximum of 55% after twelve days. Similarly, the oil density increased from 0.88 to 0.99 g/ml over this time period. After an additional four months of weathering, 10 to 15 cm size balls of mousse were noted in the tanks, along with a syrupy water-in-oil mixture that had a higher water content (and density) and a slightly lower viscosity than that observed for the discrete mousse balls. The oil/water interfacial surface tension decreased from 27 dynes/cm in the fresh oil to a value of 13 dynes/cm in the water-in-oil emulsion obtained after a twelve day period. After four months, the oil/water interfacial surface tension had decreased only slightly to a value of 12 dynes/cm. The oil/air interfacial surface tension did not change significantly over the four month period, although a very slight increase from 34 dynes/cm to 37 dynes/cm was measured. Viscosity changes from an initial crude oil viscosity of 16 centistokes to 2,800 centistokes occurred after twelve days. Four months later, the viscosity of the discrete balls of emulsified oil reached a value of 7,200 centistokes. Simple pan evaporation experiments conducted in parallel with the wave tank studies showed an increase of viscosity from approximately 26 to 100 centistokes over the time frame of day 4 through day 12.

Very high values (approaching 500 μg/l) of dissolved aromatic hydrocarbons were detected in the water column, with maximum concentrations reached at twelve hours. Due to the water column turnover (one tank volume every three hours), which was designed to simulate oil moving over previously uncontaminated water, the water column hydrocarbon concentrations decreased rapidly over the twelve day period and reached a final concentration of 10 to 20 μg/l for the total resolved components (as measured by glass capillary gas chromatography). After four months of sub-arctic weathering, and after a total volume of 2,000,000 liters of water passed beneath the slick, 5 to 15 μg/l concentrations of dissolved aromatics in the molecular weight range of methylnaphthalene

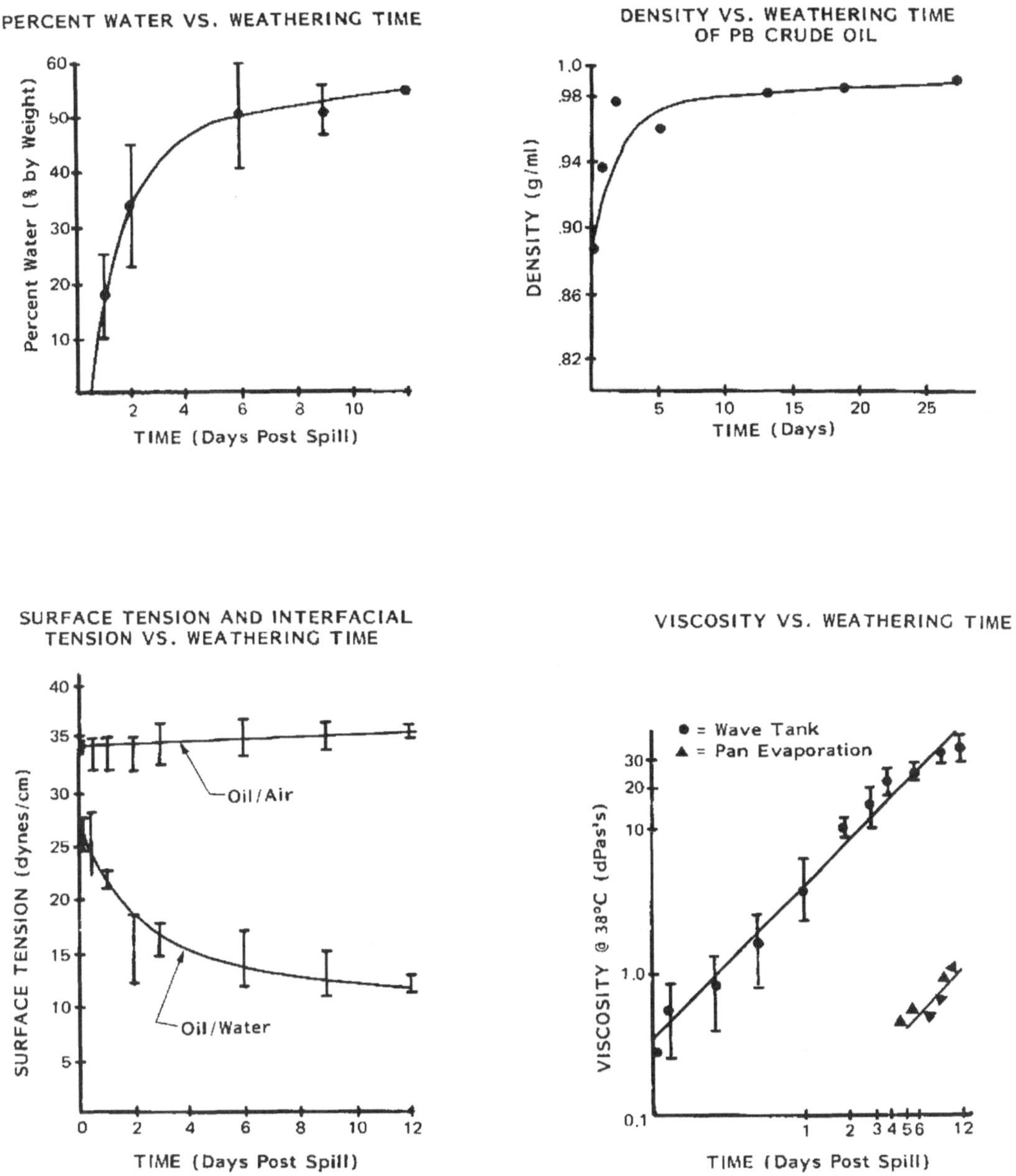

Figure 1. RHEOLOGICAL PROPERTIES DATA ON THE PRUDHOE BAY CRUDE OIL WEATHERING IN THE WAVE TANK SYSTEMS-VALUES ARE MEANS FROM THE THREE TANKS ± ONE STANDARD DEVIATION. (FROM PAYNE et al., 1983)

through pyrene were observed. The oil slick weathered rapidly during the first twelve days with components below nC-10 completely removed by evaporation and dissolution processes. After an additional four months, evaporation and dissolution weathering had removed only compounds up to nC-12. Higher molecular weight components were relatively unchanged in the mousse.

Prudhoe Bay crude oil has approximately 23% asphalts (Coleman et al., 1978), with nickel and vanadium concentrations of 13.5 ppm and 28.3 ppm, respectively. As discussed in the following section, these concentrations of surface active compounds should promote formation of stable water-in-oil emulsification; however, wave tank experiments demonstrated that this did not occur (even at 0°C) without appreciable evaporation and dissolution weathering first removing the lower molecular weight components. Even after four months the stability of the mousse was extremely temperature dependent as a melting or thawing behavior was observed when the mousse temperature was raised from 0°C to 38°C. Considerable quantities of air were entrapped in the resultant mousse, but many of these air bubbles subsequently were lost during the warming process. Nevertheless, the mixture had an extremely high viscosity (at 38°C), and additional separation of water and oil was not observed.

In several of the studies cited above, the authors noted that mousse stability was inversely proportional to the size of the water droplets entrained in the emulsions. In general, mousse with 1 to 10 micrometer diameter water droplets were most stable when they contained up to 70 to 80% water.

Although mousse can contain anywhere from 10 to 80% water, emulsions with less than 50% water generally have characteristics which, on visual examination, suggest that the oil properties are similar to that of a neat oil. These emulsions are generally free-flowing fluids with the consistency of industrial fuel oils. Laboratory studies on spreading properties of several viscous oils (with viscosities greater than 40,000 centipoise at 25°) showed that viscous oils behaved differently than non-viscous slicks (viscosities less than 500 centipoise) and that spreading of viscous oil slicks is drastically inhibited.

MOUSSE FORMATION IN THE PRESENCE OF ICE

In an effort to provide data for modeling weathering behavior of oil in the presence of ice, Payne et al. (1984b) performed a series of experiments in temperature controlled, flow-through tanks at NOAA's Kasitsna Bay, Alaska Laboratory. Through careful control of temperature, seawater flow rates, and wave turbulence conditions, ice was generated in these tanks which closely resembled (in structure and formation sequence) the first year ice characteristics reported previously by Martin (1979) and Martin et al. (1983) and observed in cores collected from first year ice in the Chukchi Sea (Payne et al., 1984b). Following the initial growth of frazil, grease, and columnar ice in seawater tanks, Prudhoe Bay crude oil was introduced beneath the ice layer, and subsequent weathering behavior of the oil throughout a simulated cycle of ice growth, thaw, and break-up in the presence of wave-induced turbulence was monitored.

One of the most important observations from these oil-in-ice studies was an extremely rapid formation of a stable water-in-oil emulsion. Emulsions of Prudhoe Bay crude oil formed within four hours of initiating turbulence (wave generation) in the presence of grease ice and breaking or rotting ice floes. The micro-scale turbulence introduced by grinding of ice floes against each other, as well as the grinding action of the frazil ice and grease ice crystals within the open water areas between larger floes, appreciably enhanced formation rates of an oil-in-ice emulsion. The grinding of ice structures also resulted in lead matrix pumping, which causes oil previously trapped beneath the ice to surface onto the ice floes and around the rims of ice floes, and enhances dispersion of oil droplets into underlying waters. This enhanced dispersion quickly becomes self-limiting, however, due to the rapid formation of a stable water-in-oil emulsion. Emulsions generated during ice breakup were not neutrally buoyant, but were sufficiently dense to cause the emulsified oil patches to reside immediately below the grease ice on the water surface. In fact, large emulsified oil patches could be observed, with adequate back lighting, just below the grease ice, near the edges of larger ice floes remaining in the tank. With continued agitation and melting of grease ice, the emulsified water-in-oil mixtures eventually surfaced into patches of open water between the individual ice floes.

Rheological properties of the test crude oil were monitored throughout the ice thaw cycle. The oil/water interfacial surface tension did not change throughout the duration of the experiment, but remained at approximate levels of 25-26 dynes/cm. Similarly, no change in the oil/air interfacial surface tension was detected during the observed emulsification process. In contrast, a uniform decrease in oil/water interfacial surface tension from 27 to 13 dynes/cm was observed during the simulated open ocean oil weathering experiments described in the previous section. Thus, in the presence of ice a stable water-in-oil emulsion was generated without the changes in the oil/water interfacial surface tension that accompanied formation of mousse under ice-free conditions. Furthermore, the water content in the emulsified oil formed in the presence of ice reached 64% within four hours after the initial ice break-up. For comparison, similar percent water contents in a stable water-in-oil emulsion under ice-free, open ocean conditions did not occur until the oil had been exposed for a period of 12 days. Large increases in viscosity were also measured in the water-in-oil emulsions formed in the presence of ice. Mousse on the ice surface had viscosities approaching 25,000 centipoise at the ambient temperature (-2°C) in the wave tank. Prudhoe Bay crude mousse at 20°C has a viscosity of 50 centipoise (Payne et al., 1984a), whereas oil contained within the brine channels of wave tank ice has a viscosity of 550 centipoise (Payne et al., 1984b). Under open ocean (ice-free) condition, viscosities of mousse may eventually reach levels of 1,100 centipoise, but only after 20 days of exposure. Consequently, these latest sets of oil weathering experiments demonstrated that mousse formation is greatly enhanced in the presence of ice, resulting in formation of stable emulsions within a period of several hours (as opposed to several weeks under ice-free conditions) that have similar water contents but very high viscosities. Rapid formation of water-in-oil emulsion has some important implications for subsequent cleanup operations. Emulsified oil is more difficult to remove from the water surface than non-emulsified oil, and is less amenable to ignition or combustion (as discussed in later sections). Furthermore, emulsions formed in the presence of ice, and subsequently transported and trapped beneath ice floes will be extremely difficult to locate from low-flying aircraft. The logistical problems of removing the emulsified oil may be too severe, and clean-up efforts feasible only after the ice melts and open-ocean methods can be initiated (e.g., Weeks and Weller, 1984; Percy and Wells, 1985).

During the Payne et al. oil-in-ice experiments, samples of the oil/water emulsion were periodically collected and subjected to analyses by gas chromatography to monitor the time-series changes in the hydrocarbon composition that accompany mousse formation. Chromatographic profiles from subsurface mousse obtained two days after ice breakup were generally similar to profiles obtained from analyses of oil which had been exposed for periods of 32 hours and 57 hours, except for the loss of some of the nC-8 through nC-12 n-alkanes. This depletion of lower molecular weight aliphatic compounds was responsible for apparent increases, relative to the fresh Prudhoe Bay crude, in percent wax (from 3.9% in fresh oil to 4.6% in mousse) and percent asphalts (from 3.7% to 7.4%, by weight). Payne et al. concluded that while the presence of surfactant materials, such as wax and asphaltenes, are necessary for mousse formation, the slight increases in their relative concentrations could not account for the observed stability of the mousse formed during the tank experiments. Instead, the rapid formation of the emulsion must be attributed to the low water temperatures (-1.7°C) and the micro-scale turbulence created by the mechanical grinding action of the grease ice crystals which injected small water droplets into the viscous oil.

IDENTIFICATION OF EMULSIFYING AGENTS RESPONSIBLE FOR MOUSSE FORMATION

Berridge et al. (1968b) found some correlation between the asphaltene content and stability of the emulsions formed. Mackay et al. (1973) attempted to determine the nature of the asphaltic compounds responsible for the stability of the mousse and, like Canevari (1969), they concluded that each water droplet was encapsulated in an envelope of surfactant type molecules with a plastic nature that prevents coalescence of the water droplets due to altered interfacial oil-water tensions.

The exact nature and identity of the surfactants responsible for this behavior has not been completely elucidated. Most laboratory studies indicate that slightly different factors are responsible for emulsion formation in different oils. In the crudes studied by Berridge et al. (1968b), the percentage of residues boiling above 370°C, the asphaltene content, and the vanadium content all showed a definite positive correlation, whereas acidity, sulfur content, and wax content showed no correlation with mousse stability. On the other hand, Bridie et al. (1980a,b) found that wax content, in combination with asphaltenes did play a significant role with dewaxed and de-asphaltenized Kuwait crude and refined lube oil; these observations are discussed in detail below. Berridge et al. (1968b) observed that water-in-oil emulsions generated with high asphaltene content crude oils were stable for many months on exposure to the elements on glass plates and on aerated seawater surfaces. These data suggested that the surface-active materials responsible for mousse formation were components of the non-volatile residues, probably asphaltenes and possibly metallo-porphyrins. When an artificial blend of Tia Juana crude was made by mixing various portions of all of the distillates except for the 30% vacuum residual, no mousse could be generated artificially, although it was stated that the metallo-porphyrins were volatile enough to have been present in the distillates. Thus, it was assumed that the asphaltenes were primarily responsible for mousse formation. These authors also reported that the presence or absence of particulate material or bacteria apparently did not affect mousse formation or the stability of mousse stranded on land or on seawater surfaces, although some evidence suggested that, under certain conditions, mousse formation can be stabilized by the presence of bacterial slime.

Zajic et al. (1974) found that a pseudomonad growing on #6 fuel oil or on aliphatic hydrocarbons formed an extracellular emulsifying agent that appeared to be a higher molecular weight polysaccharide. It was found that the emulsification behavior of this extracellular material was not affected by temperatures as low as 6°C. However, when 3% sodium chloride was added to the mixture, the emulsion broke up and a patch of surface oil was produced along with oil pellets from 1 to 2 mm in diameter. When Pseudomonad aeruginosa and two yeasts (Candida petrophilus and C. tropicalis) were grown with hexadecane as a sole carbon source, the organisms produced extracellular emulsifying agents (Friede, 1973; Guire et al., 1973). These materials were believed to contain polypeptides because they could be partially destroyed by pancreatic lipase, suggesting the presence of protein components.

Bridie et al. (1980a,b) examined the emulsion forming tendency of several crude oils. Bridie et al. found that, contrary to Berridge et al.'s results, the presence of wax and asphaltenes had appreciable effects on emulsification. A 200° topped Kuwait crude oil fraction was deasphaltized by 30-fold dilution with pentane and then de-waxed by six-fold dilution in a methylethylketone/dichloromethane mixture. Attempts to form a water-in-oil emulsion with the deasphaltized/dewaxed oil failed as the mixture lost 90% of its water after standing 15 minutes. The water-in-oil mixture from the treated oil plus the original wax exhibited a similar behavior, and the "mousse" from treated oil plus original asphaltene content lost 86% of its water content after standing 15 minutes. Thus, neither the original wax nor the original asphaltine content alone were responsible for stable mousse formation. When the treated oil plus original wax content and original asphaltene content were mixed, a stable mousse was generated. A stable mousse also formed with the treated oil plus the original wax content and only 10% of the original asphaltene content. Thus, it appeared that components in the wax were important for mousse generation, although the asphaltene fraction played a more significant role. When various lube oils were treated with 5 and 10% mixtures of the asphaltene-wax mix, stable emulsions with water contents ranging from 57 to 67% could be formed. Water droplet size in these emulsions ranged from 10 to 30 micrometers. No emulsion was formed with the lube oil blank in the absence of the added asphaltene/wax mix.

Products of photochemical and auto-oxidation have also been implicated in mousse stability. Thus, while rigid emulsions could not be formed with fresh Brega and Nigerian light crude oil (Berridge et al. 1968b), stable mousse with water contents ranging from 67 to 84% were obtained using artificially weathered Brega and Nigerian crudes. Similarly, Bocard and Gatellier (1981) examined the effects of photo-oxidation on mousse stability by irradiating a thin layer of oil (0.3mm) with a fluorescent lamp emitting ultraviolet and visible light between 300 and 450 nm with a maximum at 365 nm. In this experiment, Arabian light crude (topped at 150°C) produced oxygenated products corresponding to 0.08% after 132 hours. Using Arabian crude and Zarzatine 150°C (a crude with very little asphalt), Bocard and Gatellier found that emulsions produced with unoxidized oil generally were very unstable, whereas the emulsions made with the photo-oxidized oil were particularly stable. Thingstad and Pengerud (1983) investigated the role of photochemical oxidation, in conjunction with agitation of the oil and water mixture, in the formation of mousse from Stratfjord crude oil and seawater. Laboratory scale experiments were performed with 2 ml of Stratfjord crude in 50 ml of aged seawater in 100 ml Erlenmeyer flasks that were illuminated by visible light from a dysprosium lamp. Results from these experiments indicated that both photo-oxidation and mixing energy were critical factors responsible for mousse formation with this crude oil. No mousse formation occurred in oil and water mixtures that were agitated but not exposed to the light or in the mixtures containing beta-carotene, a known inhibitor of photo-oxidation of petroleum. In contrast, emulsions did form in the dark after addition of tetradecanal, which was considered a model product of an oxidized petroleum. Thingstad and Pengerud concluded that photo-oxidation induces formation of polar groups in the exposed oil that are soluble in the oil and act as surface-active agents facilitating incorporation of water into the oil phase.

Auto-oxidations resulting from free radical chain processes (with the rate of propagation being controlled by the rate of proton extraction from the hydrocarbon by alkyl peroxide radicals) have also been demonstrated to occur in oil-in-water mixtures. In addition to a number of oxygenated products which can be formed by these reactions, higher molecular weight polymerization products in the oil itself can result in enhanced mousse stability. In general, tertiary free radicals are found to be more stable than those from primary or secondary carbons, such that isoprenoids leading to tertiary free radicals can be more readily attacked. Alkyl-substituted aromatics such as tetralin and cumene that can be resonance stabilized are also removed rapidly by auto-oxidation. Photo-oxidation can compete with auto-oxidative processes; however, both processes are affected by the presence of vanadium and other metals of variable valence that strongly catalyze oxidations. Sulfur compounds, on the other hand, are believed to inhibit oxidation by terminating reactions caused by sulfoxide formation. All of these factors, separately and in combination, can affect mousse formation and stability. The slightly water soluble carboxylic acids, ketones, aldehydes, alcohols, sulfoxides, peroxides, etc. which are formed by these processes may serve to stabilize water-in-oil emulsions due to their surfactant type properties. However, they are also believed to be rapidly removed from the surface of the oil, leaving a more viscous, higher density residue. Once mousse formation has occurred, additional photo-oxidation of the emulsion or resultant tarry lumps must be extremely slow because of the low surface area to volume ratio. Further, photo- and auto-oxidation would tend to be limited to the external surfaces of rather viscous and diffusion-controlled emulsions where oxygen and light cannot penetrate into the interior of the mousse. As such, oxidation products would dissolve into the water column as they form. Following the IXTOC spill in the Gulf of Mexico very few oxidation products were in the mousse samples collected at the well-head during the NOAA ship Researcher cruise. It was concluded that most of the photo-oxidation products were leached from the material at or very soon after their formation.

Klein and Pilpel (1974) reported that viscous oil slicks actually appeared to contract as photo-oxidation proceeded. This was attributed to polymerization of the petroleum components and to the resulting increase in viscosity that restricted diffusion of oxidation products to the oil/water interface. In this instance, then, the authors concluded that photo-oxidation could help generate intractable tarry residues and stabilize water-in-oil emulsions. Burwood and Spears (1974) exposed surface slicks of crude petroleum to artificial light to examine the effect of dissolution of specific petroleum hydrocarbons in seawater. They postulated that indigenous auto-oxidizable hydrocarbons could react photolytically with thiacyclanes to form complex water soluble mixtures of thiocyclane oxides. Such compounds were detected in seawater following prolonged equilibrium with a medium sulfur content Middle East crude oil. The compounds appeared as a mixture of high boiling, water soluble components within an unresolved complex mixture (UCM) envelope in the nC-15 to nC-23 range of the water sample extracts. The magnitude of this UCM increased considerably as photo-degradation of the oil progressed. Burwood and Spears (1974) suggested that such a process might explain the loss of sulfur materials during weathering of crude oils at sea as occurred in the Torrey Canyon spill. A similar loss of sulfur containing hetero-aromatics, specifically benzothiophene and dibenzothiophene, was observed in the IXTOC spill where formation of sulfoxides presumably caused these materials to leach preferentially from the bulk oil before and during mousse formation.

PHYSICAL AND BEHAVIORAL PROPERTIES OF WATER-IN-OIL EMULSIONS

Evaporation

The measurable increases in viscosity and specific gravity observed for many water-in-oil emulsions in seawater affect their behavior, including spreading, dispersion, interaction with suspended particulate material, and, presumably, evaporation and dissolution properties. Twardus (1980) indicated that no quantitative data existed for mousse effects on evaporation, but suspected that once mousse formation occurred evaporation would proceed at reduced rates. Similar suggestions have been postulated by Nagata and Kondo (1977). In a field program Payne et al. (1981b) studied the quantitative loss of lower molecular weight volatile components from fresh Prudhoe Bay crude oil and from artificial mousse generated by shaker table while enclosing seawater and oil in a sealed Teflon container to prevent evaporation. Sub-arctic weathering conditions were simulated by exposing fresh crude and mousse in outdoor flow-through sea water aquariums where turbulence was induced by propeller mixing. The water and air temperatures at the time of sampling were 6° and 6 to 12°C, respectively. Figures 2 and 3 present Flame Ionization Detector-temperature programmed gas chromatograms of the volatile components released from fresh Prudhoe Bay crude oil and fresh mousse, respectively. Volatile compounds shown in these chromatograms were trapped by vacuum pumping measured volumes of air (sampled 1 to 2 inches above the slick) through stainless steel columns packed with Tenax® 1.0 and 1.5 hours after the fresh oil and mousse, respectively, were spilled on the water surface of the tanks. These traps were then capped with stainless steel Swagelok® fittings and stored at room temperature until analyses. Back-up columns, in series with the front columns, showed no break-through of lower molecular weight materials, indicating 95+% recovery on the front traps. Interestingly, the qualitative appearance of the chromatograms of the volatiles from both systems are remarkably similar. Time series data presented graphically in Figures 4A and 4B illustrate that essentially identical losses of lower molecular weight compounds, ranging from butane to xylene, were obtained for both the fresh oil and fresh mousse. The data in Figures 4C and 4D, however, show longer retention of these compounds in mousse spread on seawater in the absence of turbulence. One of the static (non-mixed) mousse systems (Figure 4D) was treated with Corexit 9527 immediately after the spill. Treatment apparently did not affect evaporative losses relative to the non-dispersant-treated control (Figure 4C). Figure 5 presents computer-generated, time series, concentration profiles from capillary FID gas chromatographic analyses of the intermediate and higher molecular weight components

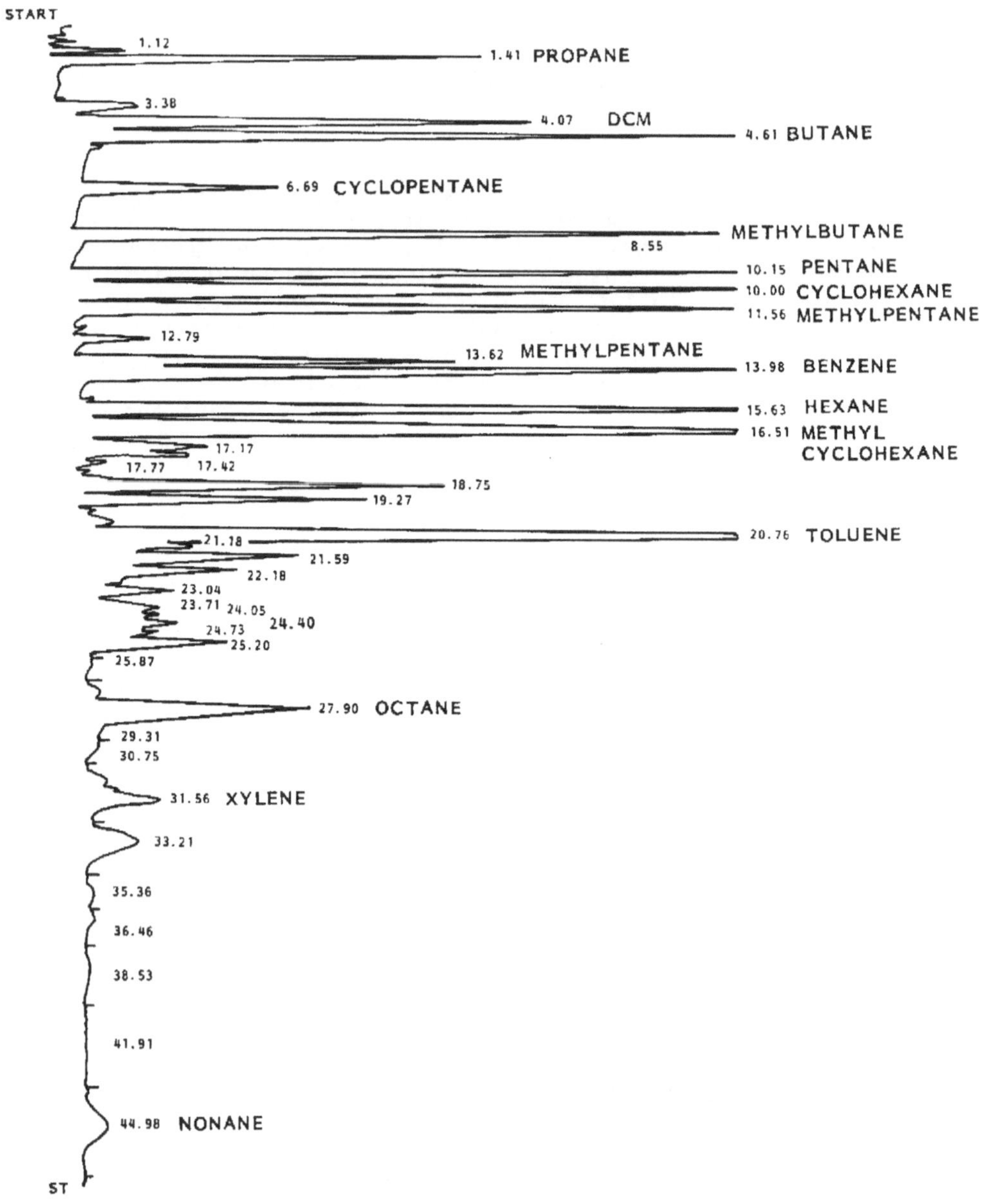

Figure 2. FLAME IONIZATION DETECTOR - TEMPERATURE PROGRAMMED GAS CHROMATOGRAPHIC ANALYSIS OF TENAX-TRAPPED VOLATIL COMPOUNDS LOST FROM FRESH PRUDHOE BAY CRUDE OIL AFTER 1 HOUR OF WEATHERING (WITH TURBULENCE) ON SEAWATER. WATER TEMP 6°C, AIR TEMP 6°C. (FROM PAYNE et al., 198lb.)

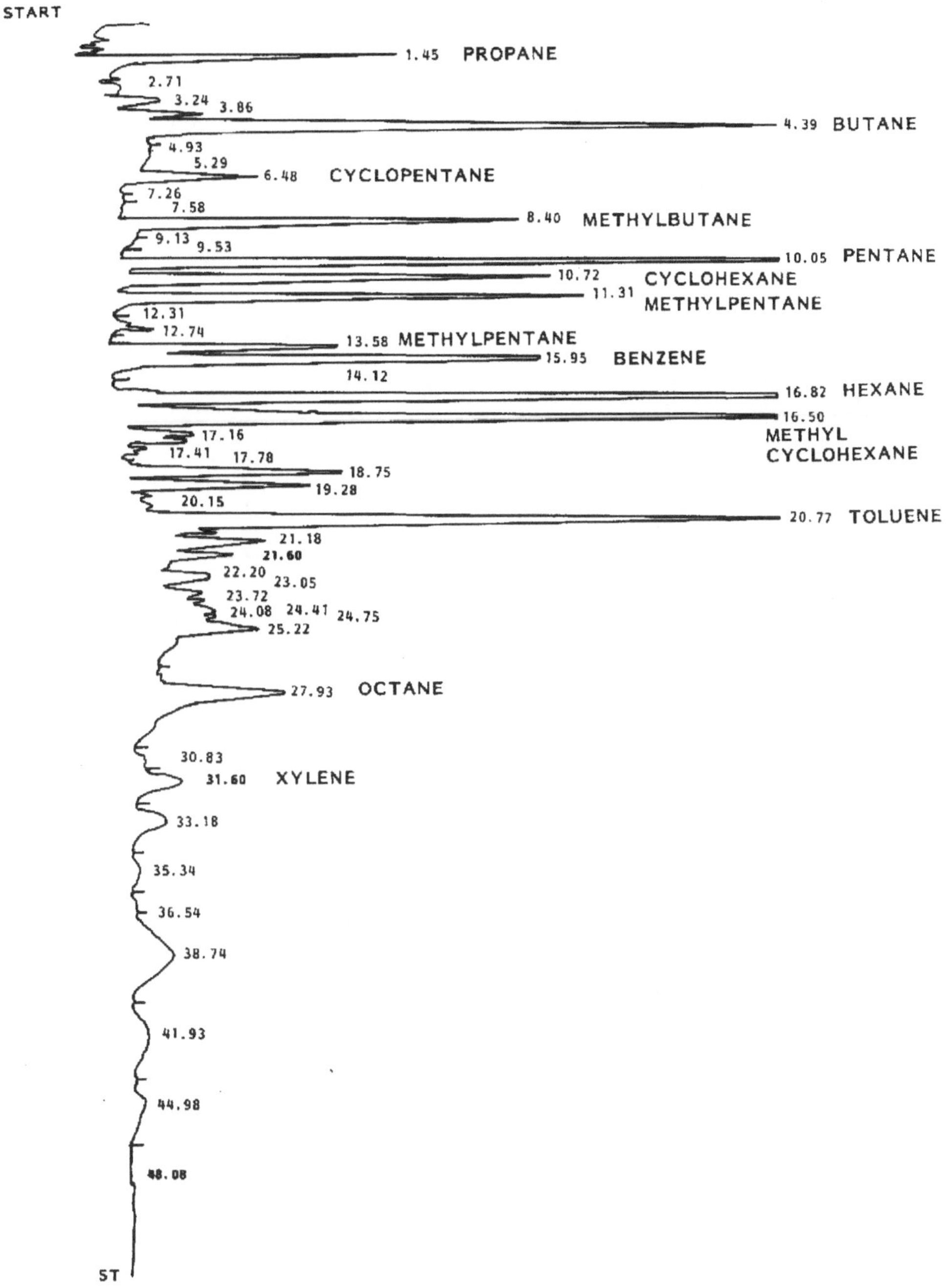

Figure 3. FLAME IONIZATION DETECTOR - TEMPERATURE PROGRAMMED GAS CHROMATOGRAPHIC ANALYSIS OF TENAX-TRAPPED VOLATILE COMPOUNDS LOST FROM FRESH PRUDHOE BAY MOUSSE (80% WATER) AFTER 1.5 HOURS OF WEATHERING (WITH TURBULENCE) ON SEAWATER (WATER TEMP 6° C, AIR TEMP 6° C). (FROM PAYNE et al., 198lb.)

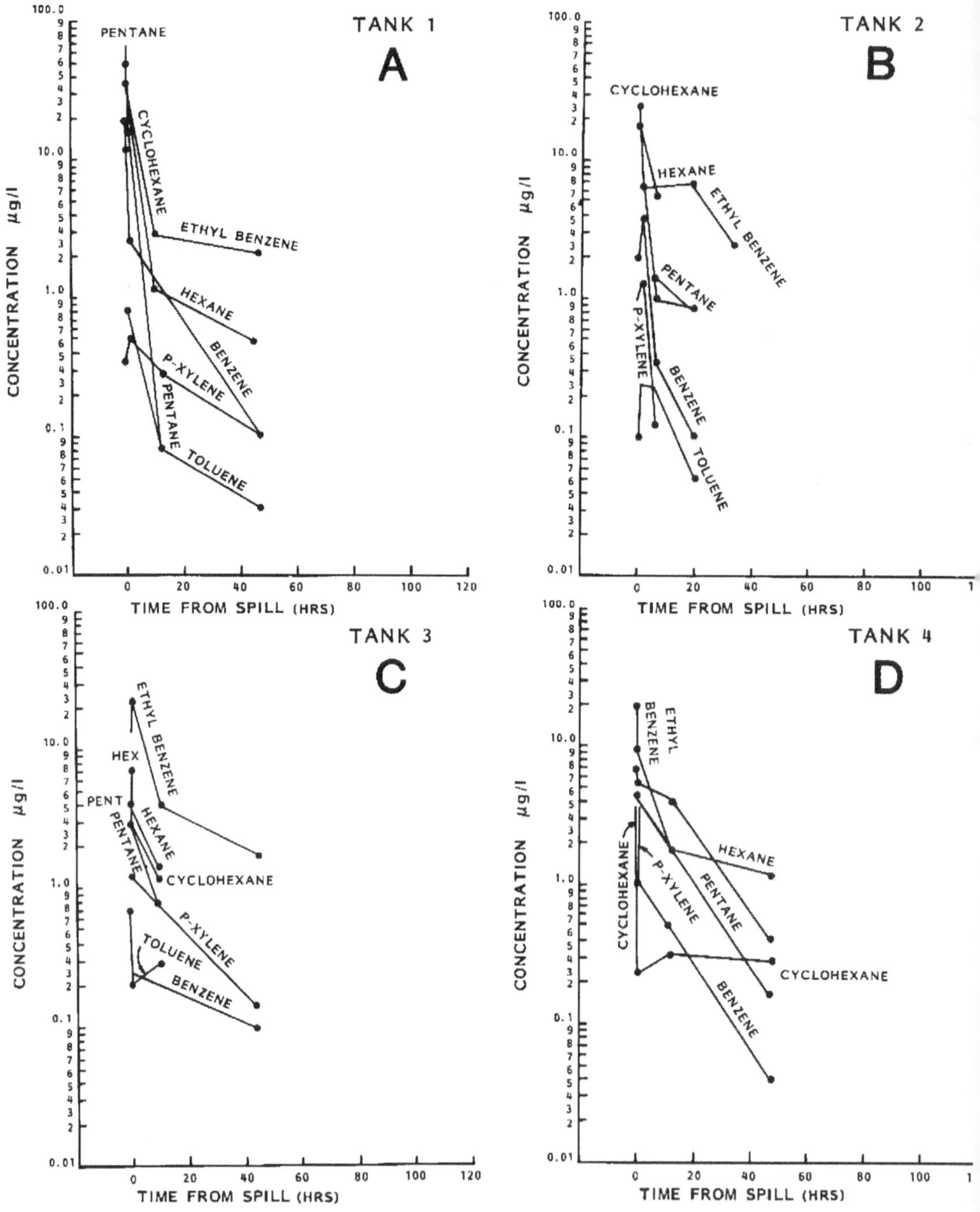

Figure 4. TENAX TRAP/FID GC DATA ON SUB-ARCTIC VOLATILE COMPONENT LOSS FROM PRUDHOE BAY CRUDE OIL AND MOUSSE ON FLOW-THROUGH SEAWATER ENCLOSURES IN KASITSNA BAY, ALASKA. A.) FRESH OIL AND TURBULENCE; B.) FRESH MOUSSE AND TURBULENCE; C.) FRESH MOUSSE (NO TURBULENCE); D.) FRESH MOUSSE AND COREXIT 9527 (NO TURBULENCE). WATER TEMP. 6°C, AIR TEMP. 6-13°C. (FROM PAYNE et al., 1981b.)

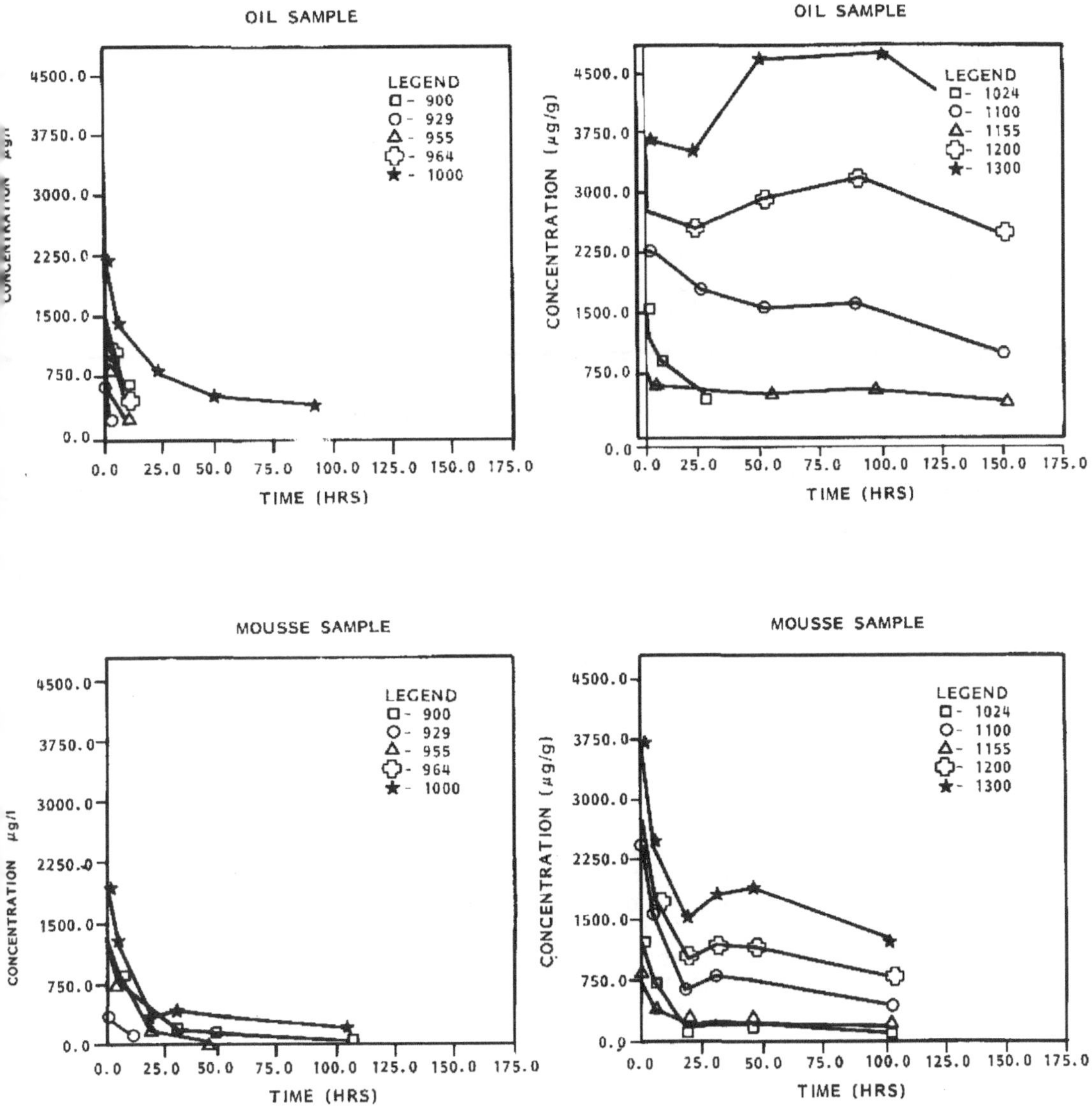

Figure 5. COMPUTER GENERATED PLOTS OF CAPILLARY FID-GC DATA ON INTERMEDIATE MOLECULAR WEIGHT COMPONENTS REMAINING IN PRUDHOE BAY CRUDE OIL AND MOUSSE WEATHERING UNDER SUB-ARCTIC CONDITIONS ON FLOW-THROUGH SEAWATER ENCLOSURES AT KASITSNA BAY, ALASKA. KOVAT INDICES ARE IDENTIFIED ON EACH PLOT. A AND B: FRESH CRUDE AND TURBULENCE; C AND D: FRESH MOUSSE AND TURBULENCE. TEMPERATURE AS IN FIGURE 4. (FROM PAYNE et al., 1981b.)

remaining in the oil and mousse from the well-stirred tanks. Kovat indices for the compounds in each plot are identified in the figures. These data illustrate that compounds in the range of nC-9 through nC-11 are retained preferentially in the mousse sample for longer time periods. Figures 6A and 6B show the time series concentrations of components with Kovat indices ranging from 1300 to 2000 for the oil and mousse samples, respectively. After 25 hours, a similar relative increase in these higher molecular weight compound concentrations (μg/g oil) is noted for both the oil and mousse. This is due to the removal of a large mass of the oil from evaporation of the lower molecular weight components. Compounds with molecular weights above nC-13 are not lost during this time.

The absolute concentrations of the individual components in each of the mousse sample plots (on a μg/g of mousse basis) are lower than those of the fresh oil because of the additional mass of the seawater (80% by weight) in the water-in-oil emulsions. Thus, in the presence of turbulence, the higher viscosity of the 80% water-in-oil mousse did not significantly affect evaporative loss of the lower molecular weight components boiling below xylene, but some reduction in evaporation was noted for intermediate molecular weight compounds (Kovat Index 800 to 1100) in the mousse.

Combustibility

Differences in the evaporation rates of volatile components from mousse and from fresh oil may affect the flash point and burn point, but more substantial changes in oil combustibility probably result from simple incorporation of water. Twardus (1980) provided the most complete characterization of the combustibility and other physical properties of aged oils and emulsions. The flash point, fire point, viscosity, and pour point were all found to increase as the percent water increased. Thus, longer pre-heating and ignition times were required for combustion of water-in-oil mousse. It was also noted that since these emulsions do not spread as rapidly, evaporation and flame propagation occur very slowly.

In water-in-oil emulsions with water contents approaching 20%, a slight elevation in fire point was observed; in heavier crudes where water concentrations exceeded 20 to 30%, fire points increased dramatically. Thus, while water-in-oil emulsions formed from fresh crude oils could be burned successfully in situ (when solid fuel ignitors were employed to initiate the combustion process), the maximum water content for effective burning of the emulsion layers was approximately 70% by volume for medium crudes and 30% by volume for heavy crudes. Twardus also noted that in emulsified crudes,

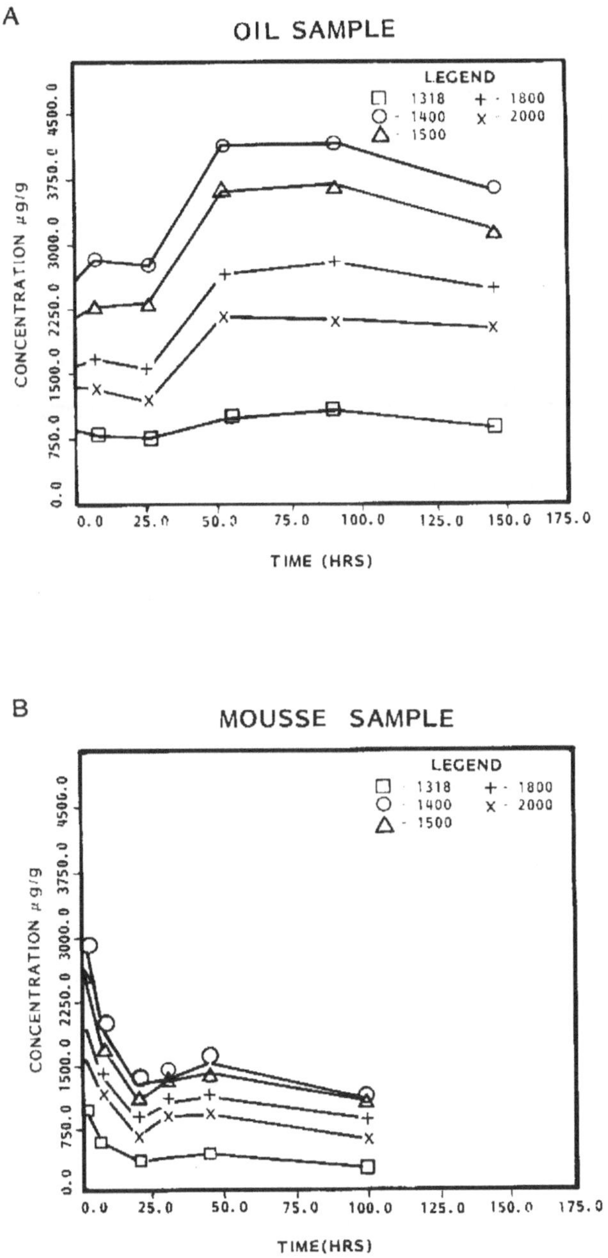

Figure 6. COMPUTER GENERATED PLOTS OF CAPILLARY FID-GC DATA ON HIGHER MOLECULAR WEIGHT COMPONENTS REMAINING IN PRUDHOE BAY CRUDE OIL AND MOUSSE WEATHERING UNDER SUB-ARCTIC CONDITIONS ON FLOW-THROUGH SEAWATER ENCLOSURES AT KASITSNA BAY, ALASKA. KOVAT INDICES ARE IDENTIFIED ON EACH PLOT, AND ENVIRONMENTAL CONDITIONS ARE AS IN FIGURE 3.

separation of water and oil must occur before effective combustion could occur. In the case of emulsified oils, this ultimately required longer pre-heating and ignition times. Further, unlike the ignition of unemulsified oil slicks, the emulsified oils required a definite size flame before flame propagation over the entire oil surface could occur.

Slick thickness of water-in-oil emulsions is also critical. For Norman Wells and Sweet Blend crude oils, having water contents up to 30% by volume, 10 mm thick slicks are easily ignited. In fact, a 10 mm thick slick of Norman Wells crude oil emulsified with up to 70% water by volume could be easily ignited with a solid fuel ignitor. Water-in-oil emulsions of heavier crudes (Lloydminster and Waburn-Midale with water contents of up to 30%) with the same thickness were much more difficult to ignite and often required the addition of fresh oil for their ignition. The residual oil layers remaining after combustion of water-in-oil emulsions ranged between 0.4 and 1.9 mm in thickness. These thicknesses were similar to residual oil layers remaining after the combustion of unemulsified oil layers of the corresponding crude oils.

Kolpack et al. (1978) reported that residues remaining after burning of the Bunker C cargo during the explosion and sinking of the Sansinena had densities considerably greater than the unburned cargo, and these residues rapidly sank to the bottom. Computer simulations of changes in composition of a standard Bunker C after two hours of evaporation at four different temperatures were generated using API standard Bunker C composition (paraffins: C-12 to C-28, with a maximum at C-14-15; naphthenes C-9 to C-25, bimodal with maximum at C-10 and C-22; aromatics C-8 to C-23 with a maximum from C-10 to C-17; and asphaltics starting at C-12 and increasing to 30% relative composition at C-48). In this simulation the density of Bünker C increased relative to that of seawater within ten hours when the oil temperature was 75°. A specific gravity of 1.025 was attained after 40 minutes at 125°C and within less than 10 minutes at 250°C. Oil simulations for unburned spilled oil showed that it would take more than one week for residual materials on the water surface to reach a density approaching that of seawater.

Interestingly, when Dickens et al. (1981) introduced an oil-gas mixture of Prudhoe Bay crude oil and air under ice, no appreciable mousse formation was observed. During spring break-up of the ice, the oil was found to be sandwiched among thin ice sheets and brine channels, but none of the oil was in an obviously emulsified water-in-oil form. This oil could be removed from the site by burning, with burn efficiencies ranging up to 95%. Final mass balance of the oil in this instance was 31% burned, 31% evaporated, 17% manually cleaned up, and 21% naturally dispersed. Nelson and Allen (1981) found little dissolution or microbial degradation of Prudhoe

Bay crude oil spilled under ice. In this case mousse formation presumably required loss of volatile components and possibly the formation of photochemical and microbial degradation products, in addition to sufficient turbulent mixing. As a result, when the oil was released in brine channels during spring breakup of ice, burning could be used as an effective clean-up/oil removal procedure. These conclusions are markedly different from those of Payne et al. (1984b) who described rapid formation of a stable emulsion with Prudhoe Bay crude oil in the presence of breaking ice (discussed above). However, in the experiments performed by Dickens et al., potentials for mousse formation may have been altered by alternate freezing and thawing of the oil-water mixtures. Dickens et al. (1981) and Twardus (1980) reported that alternate freezing and thawing cycles of water-in-oil emulsions in pans caused the emulsions to separate to some extent during aging into distinct water and oil phases. From these studies it can be see that destabilization results from freezing water-in-oil emulsions. This behavior was also observed by Payne et al. (1981b) when mousse samples were frozen.

Twardus (1980) also found that clean-up of water-in-oil emulsions was more complicated as the sorption capacity (grams of emulsion/gram of dry sorbant) of 3M brand oil sorbant was significantly reduced as the water-in-oil content increased for Lloydminster, Wayburn-Midale, Sweet Blend, and Norman Wells crude oils. Prudhoe Bay crude oil mousse generated in outdoor flow-through wave tanks at Kasitsna Bay, Alaska, also resisted clean-up by sorbants after four months of *in situ* weathering and 60 to 70% water uptake.

BREAKING AND INTERACTION OF LABORATORY MOUSSE WITH DISPERSANTS AND DEMULSIFIERS

Significant differences have been noted between the amounts of fresh oil and the amounts of fresh mousse that are dissolved and dispersed into the water column due to turbulence. For example, Table 3 presents time-series concentrations of artificially generated (shaker table) mousse and fresh Prudhoe Bay crude oil in waters within flow-through seawater systems with propeller induced turbulence (Payne et al., 1981b). The three orders of magnitude difference between the fresh oil and fresh mousse concentrations clearly reflects the latter's resistance toward dissolution and dispersion into droplets.

To evaluate the possibility of preventing mousse formation at sea, Berridge et al. (1968b) attempted to generate mousse in the laboratory with 0.1% by weight ESSO Breaxit and varying concentrations of the dispersant BP1002 added to the starting oils. Results from these and other tests are pre-

TABLE 3

Time-series water column concentrations (μg/l) of dissolved and dispersed hydrocarbons from fresh Prudhoe Bay crude oil and mousse weathering on flow-through seawater enclosures (turbulent regime) at Kasitsna Bay, Alaska. (Water temperature 6^{0} C, air temperature 6 - 13^{0} C). Concentrations determined by capillary temperature-programmed FID gas chromatography. (from Payne et al., 1981b)

Fresh Oil	0 hrs	1 hrs	7.5 hrs	26 hrs	53 hrs
Resolved Components	7200	4740	1400	10110+	659
Unresolved Complex Mixture	3140	1460	420	447	114

Fresh Mousse*	0 hrs	1 hrs	6 hrs	19 hrs	30 hrs	45 hrs	100 hrs
Resolved Components	23	7	29	18	10	24	34
Unresolved Complex Mixture	ND	ND	12	45	37	69	59

*Water column concentrations corrected for total oil volume added as "mousse."

ND = none detected

+Possibly due to excessive oil droplet entrainment

sented in Table 4. In these mixtures, the crude oil was observed to take up large quantities of water, ranging from 66% water for the Brega crude to 75% water for Gach Seran; however, no stable mousse could be formed. The time required for separation of the oil and water layers ranged from several seconds to two minutes. Photo-micrographs showed that water droplets ranging in size from 1 to 50 micrometers were encapsulated in most of the oils. However, the lowered viscosities of the mixtures (presumably from the presence of the dispersants) caused rapid phase separation. The Nigerian crude used in their study appeared to contain smaller droplets and was slightly more stable. Using ratios from 0.1 to 1.0 weight percent BP1002, various types of mousse were generated with properties ranging from stiff to sticky and soft with increasing dispersant concentration. Mousse was not formed at higher dispersant concentrations. The time required for breaking the emulsion ranged from ten minutes to two hours. Less encouraging results were obtained in attempts to break up previously stabilized mousse with demulsifers. Breaxit was tested at three temperatures and BP1002 was tested at ambient temperature. In general, relatively higher concentrations of the BP1002 were required to break the emulsion. From these results, Berridge et al. suggested that oil could be de-asphaltized before shipment, or emulsion breaking surface active agents might be added to the crude, to prevent mousse formation from large crude oil spills.

Canevari (1982) reported the results of a research and development program to formulate a demulsifier capable of separating water and oil phases. An unidentified chemical demulsifier was formulated which contained two primary components; Component A comprised a mixture of surfactants for displacing the emulsifying film which prevents water-water coalescence; Component B was a mixture of wetting agents for displacing one liquid from the surface of a bi-wetted solid with the other liquid. The chemical characteristics, as well as the ratios, of these components were critical for efficient mousse demulsification. Several candidate formulations were tested with an Arabian crude mousse containing 50% seawater by volume. The demulsifiers were added at 0.1% concentration to the mousse and then shaken to simulate a low energy mixing regime. Three of the products tested produced a separation of the oil and water phases within minutes of adding the demulsifier; separation of emulsions using two other product formulations occurred only after a settling period of several hours. Canevari also observed that demulsifier concentrations exceeding 0.2% by volume overtreated the Kuwait crude oil emulsion and produced an oily water phase, whereas concentrations of 0.01% were not sufficient for phase separation. Excessive mixing of the demulsifying agent and the mousse also produced an oily water phase. For comparison, Little (1981) predicted that relatively higher

TABLE 4

Inhibition of Stable Mousse Formation by the Addition of Chemical Dispersants (from Berridge et al., 1968b and Bridie et al., 1980a, b)

Crude	Dispersant	%	%H_2O	Appearance	Time to Separate into layers
Kuwait	Breaxit	0.1	74.5	No stable mousse	1 minute
Tia Juana	Breaxit	0.1	72.1	No stable mousse	1.5 minutes
Brega	Breaxit	0.1	66	No mousse	—
Iran (Agha Jari)	Breaxit	0.1	71.4	No stable mousse	2 seconds
Imega (Kirkuk)	Breaxit	0.1	78.1	No stable mousse	1 second
Nigerian	Breaxit	0.1	70.5	No stable mousse	1 second
Gach Saran	Breaxit	0.1	75.6	No stable mousse	2 minutes
Kuwait	BP 1002	0.1	81	Good mousse	Stable
	BP 1002	0.5	76	Fair mousse	2 hours
	BP 1002	0.75		No mousse	—
Tia Juana	BP 1002	0.75	76	Sticky mousse	10 hours
	BP 1002	1.0	65	No mousse	—
Imega (Kirkuk)	BP 1002	0.1	77	Soft mousse	12 hours
	BP 1002	0.75	67	Good mousse	10 minutes
Brent	LA 1834	0.1	-	No mousse	—
Kuwait	LA 1834	0.1		No mousse	—
Bunker C	LA 1834	0.1		No mousse	—

demulsifier concentrations (up to 0.4%) were required to separate the more stable fresh emulsions and emulsions prepared from evaporated or weathered oils within a period of several hours. However, Little observed that chemical demulsification of stable mousse would be much more difficult under arctic conditions than under tropical conditions; under cold temperature conditions additional heat and mixing energy may be required for efficient separation of the oil and water phases.

Bridie et al. (1980b) evaluated a dispersant additive (LA1834) in laboratory tests with artificial mousse generated with a Brent 200+ crude at 10°C. They found that the viscosity dropped immediately and 60% of the water separated (residual water content <50%). After the mixture stood at 30°C for four hours, additional water separated and the residual water content in the oil was less than 10%. Similar results were obtained with Kuwait 200+ mousse, but its residual water content remained higher. Adding LA1834 to either the oil or water prevented mousse formation (Table 4) at all stirring speeds and temperatures tested using Kuwait crude, Brent crude, and Bunker C fuel oil (LA1834 concentration, 300-1000 mg/ℓ). In large scale (10 ton) tank tests, addition of 3000 mg/ℓ LA1834 caused the viscosity of the 70:30 water-in-oil emulsion to drop from 130,000 cP (18°C, shear rate 3.28×10^5 g/cm S^2) to 3,800 cP; water content also decreased to 25% after ten minutes and to 15% after 60 minutes. The additive was also used successfully to lower the viscosity of recovered beached mousse from several thousand tons of Basra Light crude spilled from the Irene's Serenad, after an explosion near Greece in 1980, such that the material could be pumped from oil drums and picked up by waste disposal trucks. In this case 200 to 400g of LA1834 were added to each 200 liter drum.

In another recent study, Bocard and Gatellier (1981) studied the effects of various dispersants on water-in-oil emulsions formed from Arabian light crude topped at 150°C, (viscosity: 10,000 to 20,000 cP), a Safanya crude (viscosity: 20,000 to 50,000 cP), and heavy fuel oil (viscosity: 60,000 to 80,000 cP). To determine the effectiveness of various dispersant products, mixtures of the dispersants were added to the laboratory generated mousse with four vertical blades rotating at 500 rpm for specific time intervals. The water released as a function of stirring and settling time was measured in a graduated cylinder. Ratios of emulsion breaking product to mousse ranged between 250 to 1000 ppm; however, it was noted that above 500 ppm the gain in effectiveness was relatively slight. After mixing times in excess of 150 minutes, variable results were obtained with the seven different emulsion breaking products tested. From 18% to 56% of the water could be removed with this procedure. For the more viscous mousse, much greater agitation of the mousse and emulsion breaking product was necessary to produce similar separations. In all cases, injection of emulsion breaking product immediately caused a significant drop in the viscosity before any water separation occurred. This was illustrated for the Arabian light crude oil mousse (75% water) treated with 1000 ppm of product "A". In this case, the viscosity of the emulsion measured at 20° at the pump outlet of their laboratory emulsion breaker decreased from approximately 22,000 cP to 5,400 cP at 10 rpm after introduction of the dispersant. From infrared and TOC hydrocarbon measurements, it was also shown that approximately 20 to 25% of the de-emulsified products passed into the aqueous phase.

To determine the effects of emulsion breakers on stabilized mousse aged in a natural environment, Bocard and Gatellier performed field experiments from March 1979 to February 1980 in the oil port terminal of Antifer (Le Havre). Approximately 400 liters of Arabian light crude were poured into a floating metal enclosure producing an oil layer approximately 15 mm thick. Samples were taken periodically and the temperature was noted to vary from 6°C in winter to 19°C in summer. Up to 75 to 80% water was incorporated by the crude within the first week, and at that time approximately 30% of the crude oil components were believed to have evaporated. Over the 11 month period, water content in the mousse dropped to approximately 65 to 75% while losses due to evaporation accounted for 44% of the original oil mass. When dispersant tests were performed (with 100 ppm of Product "A"), on the samples at 20°C, a progressive stabilization of the mousse on and after September 1979 was noted, and none of the products tested were able to cause phase separation. Apparently the stability of the naturally generated mousse was not correlated directly to an increase in viscosity. The 12 month old weathered water-in-oil emulsion exhibited a 20°C viscosity of 24,000 cP at 10 rpm, 11,000 cP at 20 rpm, and 2,800 cP at 50 rpm. All of these values were considerably lower than those obtained from synthetic emulsions of Arabian light crude (topped at 150°C) with 75% water generated in the laboratory. When mixtures of the naturally generated mousse were mixed with 150°C cut Arabian light crude, an emulsion was formed that had a viscosity of only 1,600 cP at 10 rpm. This emulsion could not be broken at 20°C with any of the products tested. Similar results were obtained with other artificial mixtures of Arabian crude mousse (with 70% water) and other oils. Stabilization of the emulsions by incorporation of microparticulate particles of sedimentary material was suggested as one possible hypothesis for the observed stability. It was also stated that the formation of oxygen compounds from photochemical, microbial, and auto-oxidation processes may have acted as surfactants, with a further stabilizing effect on the emulsion.

While Bocard and Gatellier (1981) suggested that viscosity did not directly affect water-in-oil emulsion stability, slightly conflicting results were reported by Lee et al. (1981). The latter authors reported that changes in water contents and viscosities for eight crude oil mousses directly influenced dispersant efficiency. In general, it was found that dispersant efficiency with mousse mixtures having viscosities greater than 10,000 cP was very low; furthermore, little useful dispersion could be achieved above 7,500 cP for most crude oil emulsions studied. Lubricating oil base stocks gave an even lower viscosity limitation of about 2,500 cP for effective dispersant treatment. This was attributed to an absence of any natural dispersants in the lube oil stocks themselves. Other dispersant products from different manufacturers varied considerably in performance; however, they all performed better when used neat rather than diluted with seawater. Dispersants which were pre-diluted with seawater and applied to the oil were usually less effective relative to results obtained with undiluted dispersants. Dispersant effectiveness was affected by viscosity when viscosity of the oil exceeded a limiting value, and the dispersant was removed physically from the oil by wave action before it could diffuse into the slick. In independent investigations, Mackay et al. (1980) stated that viscosity increases in mousse are critical in dispersant treatment, and that viscosity can be correlated mathematically with percent water uptake. In their studies they concluded that mousse with viscosities in excess of 4,000 centipoise would be difficult, if not impossible, to disperse.

Dispersants have been used with mixed results in various major open ocean and coastal spills. In the Amoco Cadiz spill, when emulsion breakers were used in intermediate storage tanks and mobile van trucks, very little separation of water from the mousse occurred. The addition of dispersant did, however, speed up transfer of the mousse by reducing the viscosity of the emulsion. Bocard and Gatellier (1981) hypothesized that mixing energy in the trucks or storage tanks was not sufficient to release a large proportion of the water. Recommendations were made during clean-up operations to inject emulsion breakers in-line, upstream from the pump at a flow-rate that was proportional to pumping throughput. Also, in field applications, when emulsion breakers were applied to slicks, the efficiency of skimmers improved. However, preliminary tests performed in the laboratory with slight stirring have shown that the emulsion breaking action is very slow under these conditions.

BACTERIAL UTILIZATION OF LABORATORY GENERATED MOUSSE

As noted earlier, results reported by Berridge et al. (1968b) indicated only minimal microbial utilization of laboratory generated mousse. Likewise, only limited evidence of bacterial degradation of 350° topped Kuwait crude oil mousse was observed by Davis and Gibbs (1975). Some evidence of n-alkane removal (presumably due to biological activity) was observed in their open tank system, although this was not observed in the closed tank system. In the closed tank system microbial degradation was believed to be limited over time by dissolved oxygen and/or nutrient depletion. Oxygen and nutrient concentrations in the surrounding bay waters were such that microbial processes were not affected in the bay or open tank. It was clear from their study, however, that truly effective degradation would occur only after physical dispersion of the mousse into smaller particles. In outdoor, flow-through, wave-tanks located at Kasitsna Bay (Lower Cook Inlet) Alaska, Payne et al. (1984a) found little evidence of microbial degradation of Prudhoe Bay crude oil mousse over a four month (July-October, 1982) period. Similarly, Nagata and Kondo (1977) reported only limited microbial degradation occurred on laboratory generated crude-oil mousse.

In open ocean oil spills, variable rates of microbial degradation and utilization of components in mousse have been observed. In the Amoco Cadiz spill, for example, there was abundant evidence indicating appreciable weathering via microbial degradation processes. Some investigators suggested that microbial degradation rates even approached losses due to evaporation and dissolution. On the other hand, in the IXTOC and Potomac spills there was little evidence of microbial degradation of the mousse or tar flakes. These phenomena will be discussed in greater detail in the next chapter on case studies of mousse formation from major open ocean and coastal spills and blowouts. After the Tanyo oil spill, mousse samples were collected and oxygen consumption was measured after adding mixed microbial cultures to mousse/water mixtures; however, no activity was detected during the first 200 hours. When the sample was treated first with a surfactant (1% by weight) a positive reaction was observed and the level of bacterial activity was related directly to the increased surface area and amount of oil available for degradation (Bocard and Gatellier, 1981).

CHAPTER 3

SELECTED CASE HISTORIES OF THE MORE DETAILED CHEMISTRY STUDIES OF MOUSSE BEHAVIOR AND LONG TERM FATE IN NEAR-COASTAL AND OPEN OCEAN OIL SPILLS/BLOWOUTS

A number of fairly large oil pollution incidents which involved mousse formation have occurred during the last decade. Table 5 presents details from several selected cases where mousse formation was noted and, in some cases, where chemical characteristics of the mousse were studied. In the following summaries, only those factors dealing with mousse formation are included. There are obviously some gaps in our knowledge and details for some of the oil spills are incomplete. Nevertheless, because of the number of spills and scientific investigations during the last several years, it is not surprising that the most detailed chemistry data have only recently become available. Thus, most of the deficiencies in our data base are from the earlier spills; these spills and numerous other spills which have occurred over the last 15 years are not included in this review. In many cases these earlier spill events were not studied in detail, or if they were, the physical/chemical data are not available at this time.

It will be noted from examination of Table 5 that most of the oils which generated mousse in real spill events also formed stable water-in-oil emulsions in laboratory experiments (see Tables 1 and 2). Very limited data were available from spills where mousse formation did not occur. As a result, relatively fewer field data on oil spills involving other crudes are available. Nevertheless, it is important that the majority of oils which have been shown to form mousse in laboratory studies also cause the most severe problems with mousse formation in actual oil spills, as reported in the literature.

TORREY CANYON

The *Torrey Canyon* spill occurred in March 1967 near the Cornwell coast of England. This was the first major spill event in which mousse formation received notable attention in terms of transport and clean-up. The Kuwait crude oil transported by this vessel reportedly formed a semi-solid, gel-like mousse following the spill at sea and during clean-up operations (Smith, 1968). These emulsions were fairly stable and some contained up to 80% water. In this instance, the spilled oil was fairly heavy with only 21% of the composition-percent distilled at 210°C. The oil composition consisted of 31% saturates, 33.7% aromatics, and had an

Spill/Blowout Incident	Location	Date	Oil Type	Volume (tons) *	Water Temp	Air Temp	Mousse Type	Fresh Oil Composition						Oil/Mousse Properties					Reference
								SG.	Aliphatic	Aromatic	Polar	Asphaltene	Visc./Pour Pt.(°C)	H_2O%	Asph	Density	Visc/Pour Pt.	H_2O Droplet Size	
Torrey Canyon	W. Cornwall, England	March, 1967	Kuwait crude	~ 100,000			Semi-solids Grease-like	0.869	31.1	33.7			12°	50-80					Twardus 1980, Berridge 1968b Mackay et al., 1973
Arrow	Chedabucto Bay, Nova Scotia	Feb., 1970	Bunker C	12,000	0-2°		Grease-like	0.950	28**	25**	29**	20**	19,600cP/-1°	40-60					Owens 1978, Rashid, 1974 Mackay et al., 1973
Metula	Straights of Magellan	Aug., 1974	Kuwait crude Bunker C	52,000 2-3,000	8-10°	~ 8°	Extremely stable Light Brown/Dark Brown	(See Table 1 for examples)						25-30 5-10					Straughn, 1977; Hann, 1977
Boehlen	Brittany, France	1976	Venzuela crude	10,000				1.0	2	36			10,000cst/12°						Maurin, 1980
Ekofisk Bravo	North Sea, Norway	April, 1977	Ekofisk crude		6°		Yellowish/brown-unstable; after well capped more stable brown mousse observed (more weathered)	0.844	42	21			7,500cP/6°	83 50-60	0.95		70,000cS/6°	1 um	Cormack and Nichols, 1977, Grahl-Nielson, 1977 Audunson, 1977
Potomac	Melville Bay, Greenland	Aug., 1977	Bunker C Pitch 55%	308	3-4°	4°	Oil globs(pancakes) & flakes; no apparent emulsification	0.96 1.055						≤5					Petersen, 1977
Amoco Cadiz	Brittany, France	March, 1978	Arabian crude & Bunker C	230,000 4,000			Reddish-brown Quite stable	0.853	39	34	24	3	4-10cSt/	>50-80 variable		0.98-1.028			Calder & Boehm, 1981; NOAA/EPA Special Report in 1978, Hann et al., 1978
IXTOC I	Bay of Campeche Gulf of Mexico	June, 1979 March, 1980	IXTOC crude	476,000	25-28°	20-33°	Brown/stable, Delayed formation until after evap & hν weathering	0.883	52	38	7-8			60-70		0.99			
Burma Agate	Galveston, TX	Nov., 1979	Nigerian crude	263,000 bbl	-	20-33°	No mousse formation noted	0.828											Kelly et al., 1981
Planned Spills																			
	East Coast	Oct/Nov, 1975 & 1978/ 1979 (with dispersants	Murban crude (Middle East)	440 gal	11-14°	12-17°	Delayed formation of reddish/brown mousse	0.830											JBF/API, 1976; McAuliffe et al., 1981.
			La Rosa crude (Venzuela)	440 gal			No mousse noted but thicker "lenses" observed at leading edge of slick.	0.910											
	West Coast	Sept., 1979 (+/- dis-persants)	Prudhoe Bay (Alaska)	10-20 bbl			No mousse formation noted but thicker "lenses" observed at leading edge of slick.	0.89					183SSU			0.92	1210SSU		McAuliffe et al., 1981
Alvenus	Galveston, TX	July, 1984	Venezuelan crudes	45,000 bbl	30°	29°	Surface slicks contained viscous tar balls with consistency of roofing materials.	.857 (Merey)	23	7.8			230cP	33		0.94			Payne and McNabb (unpublished data)
								.895 (Pilon)	16	11			2,200 cP						

* Volume in tons except where noted
** Composition based on std Bunker C oil

initial density of 0.866 g/ml. There were several discussions of oil spill clean-up problems of the beached mousse along the Cornwall coast; however, most of the reports centered on the toxic effects of the dispersants used in the oil spill clean-up. These effects received much greater attention in the literature immediately following the spill.

TANKER ARROW

The Liberian Tanker Arrow carrying 108,000 barrels of Bunker C crude oil grounded off Cerberus rock in Chedabucto Bay, Nova Scotia in February 1970. It was estimated that approximately half of the cargo was spilled. Portions of the cargo were transported out to sea, although a considerable fraction also moved onto the coast, contaminating approximately half of the 600 km shoreline (Keizer et al., 1978). Mackay et al. (1973) reported that much of this oil was in the form of mousse containing up to 40% water. It was also noted that because of the high viscosity of the cargo and the cold water temperature (0-2°C), the oil was observed (by divers) to be released as "discrete pieces, like a rope 1-3 feet long" through holes in the hull of the vessel (Barber, 1970). While much of this very viscous oil and mousse eventually reached the shores of Chedabucto Bay, large portions were also broken up into smaller particles during the first 15 days following the spill by heavy turbulence and by the addition of ten tons of the dispersant Corexit 8666. These particles were detected at considerable distances from the vessel. Forrester (1971) reported that particles of finely dispersed oil, ranging from 5 mm to 10 mm, and occasionally as large as 2 cm, were found in the water column to a depth of 80 m. In general, the total oil concentration decreased with depth; however, the relative abundance of the smaller particles increased. Finely dispersed particles in a band extending up to 25 km offshore could be traced from the vessel to a site approximately 250 km southward from Chedabucto Bay. Two weeks after the wreck, particles were still observed 70 km to the east of Nova Scotia in 10 km wide tongues along the surface.

Conover (1971) reported that large quantities of the finely dispersed oil from the Arrow were ingested by zooplankton. It was believed that most of this ingested oil was eliminated in the feces which were observed to contain up to 7% oil by weight. Conover estimated that as much as 20% of the oil droplets with diameters less than 0.1 mm could be removed from the water column and incorporated into feces which rapidly sink because their specific gravity exceeds that of seawater. In a related study, Parker (1970) found oil droplets in the gut contents and fecal pellets of copepods and barnacle larvae. In both instances, there was little evidence to suggest that any adverse effects to the organisms occurred as a result of the ingested oil. Parker et al. (1971) later estimated that one copepod (Calanus finmarchicus) conceivably could ingest up to 1.5 x 10 grams of oil per day. Thus, a population of 2,000 individuals per cubic meter of seawater ingesting oil at this rate, and covering an area of one sqare kilometer to a depth of ten meters, theoretically could remove as much as three tons of oil per day if oil concentrations were 1.5 micrograms per liter or greater.

With regard to the stranded mousse from the Arrow spill, Thomas (1977) stated that in 1970 most shoreline sampling stations showed 100% oil cover at mean high water. Surface coverage at the lower tidal levels decreased with time at a logarithmic rate, with shores exposed to heavy wave and ice action showing the fastest removal. Owens (1978) also reported that the residence time of stranded oil increased as the degree of energy in the shorelines decreased. By 1973, most of the oil in the lower cliff zone (lower 27 to 33% of the tidal range) had been removed. Above this level, the percentage of surface oil coverage increased to the mean high water mark and then rapidly declined at higher elevations. Slow removal occurred at higher tide levels and sheltered locations; under the calmer conditions in the sheltered areas, surface oil existed at the lower tidal levels until 1973. By 1977, surface oils persisted in the more sheltered areas and the extent of oiling was directly proportional to the wave energy and degree of sheltering of the stations. Correlations of oil content with intertidal and subtidal substrates were noted, with long-term contamination increasing from: 1) broken rock and boulder, to 2) bedrock with sand at the high water level, to 3) broken rock and gravel, to 4) muddy sand at high water, and to 5) sandy mud at low water.

Toxicity to the salt marsh cord grass (Spartina alterniflora) did not occur for a one year period because at the time of the initial oiling the plants were dormant and the aerial portions were not coated directly with the oil. Furthermore, the oil could not penetrate into the soil as the ground was frozen and the oil was extremely viscous. In the spring of 1970, the marsh grass penetrated the oil and foliage appeared normal. During the following summer, however, the oil was remobilized during warm periods, and during the following year the number of plants was reduced and some of the plants were chlorotic.

Rashid (1974) reported on changes in Arrow oil viscosity as a function of weathering conditions. In a sample of the cargo obtained at the time of the spill, aliphatic and aromatic hydrocarbons constituted 51% of the total, asphaltenes made up 20%, and resins and NSO's contributed 29%. The viscosity of the stored cargo was 19,584 cP. The viscosity of stranded oil samples collected 3.5 years following the spill ranged from 28,600 cP to 1,210,000 cP to 3,640,000 cP in low energy, moderate energy, and high energy coastlines, respectively. At a site protected from wave action at all times, the oil was observed to have a low viscosity and a high total hydrocarbon content (49%). Thus, oil on beaches exposed to continual wave action weathered to mixtures having higher viscosities and increased asphaltene contents relative to the starting crude.

In a subsequent study where oil composition was considered seven years after the Arrow incident, Vandermeulen et al. (1977) observed that aromatic and cycloalkane components were significantly more resistant and unaltered compared to the aliphatic components. The site considered in their study was part of a gently sloping shoreline of a large low energy lagoon system. The site contained a 2 to 3 cm thick tar layer up to a meter wide lying along the high tide line at the top of the beach. Oil concentrations in the sediments ranged from 6.7 to 5,500 ppm. Increasing concentrations with depth were noted in the high tide sediments, but the reverse was observed in the low tide sediments. Significant degradation of the aliphatic components was noted; the unweathered Bunker C contained a normal compliment of n-alkanes from nC-13 through nC-30, whereas the weathered samples showed almost total loss of alkanes up to nC-30. This was most pronounced in the mid- and low-tide samples, although no differences in alkane losses were noted with depth in the sediments. Large unresolved complex mixtures characterized the gas chromatograms of the unweathered Arrow Bunker C and of all of the weathered samples. Synchronous fluorescence spectra of the original oil and extracts from the high-, mid-, and low-tide zones appeared to be essentially identical, illustrating the extreme longevity of the polynuclear aromatic components in sediments. The loss of the alkanes and the predominance of the aromatics was generally attributed to microbial activity.

Keizer et al. (1978) reported that an oil and sediment mixture having a "pavement like" consistency was found in the upper intertidal zones of Rabbit, Crichton, and Durelle Islands six years after the spill. Evidence of oil contamination was also found in the intertidal sediment in many other areas of the Bay; however, the pavement-like material was limited primarily to the more sheltered locations. Along portions of the shores of Rabbit Island and Inhabitants Bay, oil, sand, and pavement mixtures up to 15 cm thick were noted in the upper half of the intertidal zone. This oil was observed to spread out and flow on hot days causing additional leached components to enter the interstitial waters. During these studies there was also evidence of more recent oil sedimentation on the beach, so they were not able to evaluate completely the overall extent of weathering and mobility of stranded Bunker C from the Arrow. Thus, while most of the oil stranded on the shores of Chedabucto Bay disappeared over the six year period, there was evidence of persistent contamination in many locations. Specific identification of much of this oil was impossible, but a few isolated areas had visible oil contamination which could be identified as Arrow Bunker C.

METULA

The Metula grounded off of Satellite Patch, west of the first narrows in the Straits of Magellan on August 9, 1974. From that date until 25 September 1974, 50,000 to 56,000 tons of oil were spilled. Most of this was Kuwait crude, but 3,000 to 4,000 tons of Bunker C were also lost after the first few days of the grounding (Straughan, 1977). This spill occurred during the southern hemisphere winter and spill transport and weathering were affected by extreme turbulence and very cold conditions (water temperatures during the surveys ranged from 8-10°C while air temperatures averaged 8°C) (Straughan, 1977). It is not known whether mousse formed immediately upon release of the oil to the sea, but two distinct types of mousse were observed during field studies two weeks later. A dark brown mousse containing 5 to 10% water presumably was generated from the Bunker C fuel, whereas the light Arabian crude formed a light-brown mousse with 25 to 30% water (Hann, 1977). A total of 25 miles of coastline, including two small tidal estuaries, were heavily impacted during the first three weeks following this spill. Both types of mousse were deposited high on the beach front by spring tides and waves, and both types of mousse had incorporated seaweed, sand, and numerous small organisms. The darker mousse ranged from 5 to 10 cm thick and covered from 6 to 15 meters of flat areas at the top of the beach. The lighter material typically covered from 15 to 60 meters of beach at depths from 1 to 5 cm. During the first site visit, between 75 to 90% of the total oil washed ashore along the 40 mile stretch of Tierra del Fuego. Five months later, between 120 and 150 miles of beach were visibly impacted, and at several locations dozens of black tar balls ranging from 3 to 5 cm in diameter were observed. These consisted primarily of weathered oil, shell fragments and sand. In other areas, tar balls ranging from 0.5 to 8 cm in diameter occurred in beach sands. Mud flats were perhaps the most severely impacted, containing bands of mousse and sand mixtures from 15 to 25 meters wide. Like the mousse-sand mixtures from the Arrow incident, this material had hardened during the southern hemisphere summer like a road or sidewalk. In January 1976, the lower tidal zone at Puerto Espora contained considerable amounts of mousse and sand-asphalt material. At this site the intertidal zone is from 500 to 600 meters wide and about 3 to 4 kilometers long. Interestingly, the seaward 100 meter wide stretch of the lower intertidal zone (exposed only at low spring tides) generally was not coated with oil (Straughan, 1977). However, at the higher elevation in the lower intertidal zone, approximately 60% of the western third, 95% of the middle third, and 80% of the eastern third was covered with black asphalt-like material. This mousse ranged from 5 to 10 cm thick, but reached 15 cm in some areas. At one area, mousse mixed with sand was noted to a depth of 30 cm. Heavily impacted areas will obviously take years to recover and will be a continuous source of contaminants to interstitial waters and to the general coastal environment.

EKOFISK BRAVO BLOWOUT

The Ekofisk Bravo well blowout occurred on April 22, 1977, and was the first major blowout in the North Sea area. In perhaps the fastest response time to any spill event to date, a detailed chemical, biological, and physical oceanographical study was initiated within 36 hours following the blowout (Grahl-Nielsen, 1978). Also, before the blowout, Cormack and Nichols (1977) had conducted a number of field tests with 0.5 ton amounts of Ekofisk oil, thus additional information on weathering behavior in the event of a major spill was available. They had shown that while emulsion formation was as rapid as for Kuwait crude oil, the resulting viscosity of the Ekofisk emulsions was low and insufficient to allow interference with the natural spreading and dispersion rates. Evaporation was a primary removal mechanism for the Ekofisk oil, as some 53% of the weight of the oil boils below 350°C. During the test spills (sea temperature 11.6°C, air temperature 18°C, wind speed 12 knots with gusts to 18 knots, and sea state 3-4 on the Beaufort scale) up to 21% of the weight of the oil was lost to evaporation in 7.5 hours. After three hours, there was no evidence of any hydrocarbons below nC-9, and after 7.5 hours hydrocarbons up to C-11 were lost and C-12 and C-13 hydrocarbons were substantially depleted. Thus, while emulsification did occur in the test spills, it did not have a serious effect on the rate of evaporative loss.

Ekofisk crude has a relatively low asphaltene content (0.03% by weight) when compared to other oils such as Kuwait crude (1.45% by weight), that are known to form stable water-in-oil emulsions. Therefore, extensive stable mousse formation was not anticipated with the Ekofisk Bravo oil. Cormark and Nichols did observe that the Ekofisk crude oil took up water from 35% to 72% in a 0.5 to 1.3 hour period following the test spills. However, the emulsion rapidly broke up into patches of about 1 to 5 cm in diameter. After approximately 21 hours, these patches were further broken down into small 5 to 10 mm diameter flakes. It was also noted that the rate of water-in-oil emulsification was extremely dependent on the degree of turbulance and sea state. At wind speeds in excess of 12 knots the oil incorporated 70 to 80% water in less than 2 hours, although at 2 to 3 knots the rate was 10 to 20 times slower. Using these data, Mackay et al. (1980) estimated half-times for emulsion formation of 2.8 hours at 3.1 knots, 16 minutes at 10 knots, and 1.6 minutes at 31 knots. Thus, oil behavior is extremely dependent on physical oceanographic conditions. Mackay et al. (1980) stated that a very real need exists to determine which of the dispersion and emulsion formation processes (which are competitive) occurs with various oils under given temperature and sea state conditions.

In the actual spill situation from the Ekofisk Bravo rig, Grahl-Nielsen (1978) measured levels of oil-in-water at approximately 100 μg/l to 400 μg/l. These values clearly suggested the presence of oil-in-water emulsions. Grahl-Nielsen used GC/MS analyses to characterize naphthalene, phenanthrene, and dibenzothiophene (NPD) components in the oils. He also reported three basic observable phases which occurred during and after the blowout; first, during and immediately after the blowout, fresh oil appeared on the surface and at depth as an oil-in-water emulsion; second, two weeks after the blowout was stopped, the remaining oil was found in small lumps north of the wellhead; and third, 4 to 5 weeks after the blowout was capped, small oil patches on the surface of the sea were observed south of the wellhead. Although an estimated total of 20,000 tons of oil were eventually released, no napthalene, phenanthrene, or dibenzothiophenes could be detected in the water column under the oil patches isolated 4 to 6 weeks after the blowout. At the time of the blowout, these components totaled approximately 8 μg/liter in the water near the wellhead, whereas their concentrations dropped to 0.1 μg/liter outside the immediate well vicinity. Two weeks after the blowout was stopped, the NPD concentration dropped to 0.4 μg/liter near the well and to 0.05 μg/l (the limit of detection) with increased distance from the wellhead. No depth gradient was noted at that time.

Audunson (1978) also studied the fate and weathering of the surface oil from the Bravo blowout. The oil, at a temperature of 75°C, was sprayed into the air, and an estimated 35 to 40% of total material were lost by evaporation before and after the oil hit the water surface at 6°C. The specific gravity of the fresh crude was 0.85 to 0.87, but increased to 0.95 after two days of weathering on the sea surface. The oil was observed to spread in a 1 mm thick slick in 100 to 200 m wide bands up to one kilometer long. At the outer edge of these bands, 1 to 20 mm thick water-in-oil emulsions with a yellowish-brown color and up to 70% water were noted. These emulsions were very unstable and rapidly broke into long strips 1 to 20 mm thick and ten meters wide. Bands of oil were further broken up by turbulence. A more stable mousse, formed around the time the well was capped, had a brownish appearance and contained up to 50 to 60% water. This mousse did not have the rigid structure of a Kuwait crude oil mousse, and broke up into smaller 2 to 20 mm patches and, ultimately, 1 to 3 mm droplets. Microscopic examinations of the 56% water-in-oil emulsion showed 1 µm sized water droplets to be dispersed in the mixture. The slightly more stable mousse which formed after the well had been capped presumably was stabilized partially by bacterial or photo-oxidation products that were not present in starting crude. The viscosity of the oil on the sea surface increased from an initial value of approximately 1,500 cP (at 6°C) to a value of 76,000 cP (at 6°C) after two days. This increase was attributed to weathering and evaporation plus incorporation of water (Audunson, 1978). Even this relatively more viscous mousse, however, rapidly broke up into smaller patches.

To determine if the dispersed Ekofisk oil was deposited in the sediments near the well site, Johnson et al. (1978) undertook a detailed sediment sampling and analyses program. GC/MS mass fragmatograms, nC-18/nC-19, and nC-27/nC-26 ratios were used, along with hopane diastereomer ratios, to characterize Ekofisk Bravo crude from background pollution. Even immediately after the blowout, Ekofisk Bravo oil levels in bottom sediments were low relative to hydrocarbons from other materials and anthropogenic sources. Four to six weeks after the spill the hydrocarbon levels in the sediments surrounding the wellhead had returned to background levels. The majority of sediment samples contained levels of Bravo oil less than 1 ppm, and the maximum observed levels were all less than 8 ppm. All the sediment samples were collected with a Smith-MacIntyre grab, however, and the low oil levels may be due, in part, to the loss of surface flocculant material during the sampling. This phenomenon will be discussed in greater detail later when considering the IXTOC spill in the Gulf of Mexico.

Addy et al. (1978) looked for changes in the biological populations occurring near the wellhead. Despite the fact that changes were noted in the biological community, it was not possible to distinguish between the effects of the spill and effects from other, unrelated activities. Chronic petroleum pollution in this area, sediment disturbances from anchoring, and pipeline installations were believed to have caused as great a perburbation as the well blowout to the biological species present. From this study it was not possible to attribute any specific changes to the Ekofisk Bravo incident. Thus, the overall impact of the Bravo spill was relatively minor because no coastlines were impacted. The absence of significant environmental effects exemplifies the differences in overall environmental impact between open ocean versus coastal spills.

US/NS POTOMAC IN MELVILLE BAY, GREENLAND

On August 5, 1977 the US/NS Potomac was holed by an iceberg while being escorted by the USCGC Westwind through scattered sea ice in Melville Bay in the northeastern part of Baffin Bay off Greenland during intermittent dense fog. Approximately 380 tons of cargo (primarily Bunker C crude oil, specific gravity 0.96, containing 55% pitch, specific gravity 1.055) were lost. Petersen (1978) studied the oil weathering behavior shortly after the spill. Evaporation and dissolution were the primary weathering mechanisms operating following the spill, although the low water temperatures (3-4°C), light winds (0-7 knots), and thick oil slick (up to 0.75 cm) all contributed to low evaporation rates. Nevertheless, alkanes up to nC-17 and substituted naphthalenes were depleted by as much as 50 to 100% after 15 days. Shortly after the spill the oil was seen to form small pancakes (10 to 20 cm in diameter and 0.5 to 0.75 cm thick), organized in wind rows about four meters wide. Sheen (visible thin slick) was observed to emanate from the pancakes during the first two weeks. Calm seas (waves 0 to 2 feet) prevented any appreciable dispersion for the first several days, and by August 19, 14 days after the spill, pancakes ranging from 8 to 15 cm in diameter were still in wind rows several hundred meters long and up to seven meters wide. By August 20, 80% of the pancakes no longer produced a sheen, and the majority of the mass of pancakes was submerged. In addition to the pancake phenomenon, flake-like particles of tar were observed at the water surface and in the water column ten days after the spill. These were possibly from submerged oil droplets which had re-surfaced after dissolution of water soluble components. The surface oil was spongy in texture, although it did not form a water-in-oil emulsion. Even after two weeks of weathering, less than 5% water was found in the surface collected oil.

Relative increases in methyl phenanthrenes in the weathered oil and in sub-surface flake samples suggested that enhanced dissolution of the lower aromatics may have occurred, although gravimetric measurements of the asphaltenes showed no significant changes from earlier (5-10 August) samples. The absence of any change in nC-17/pristane and nC-18/phytane ratios over a two month period suggested that appreciable microbial degradation of the oil had not occurred. It was estimated that most of the residues eventually sank and reached the bottom of Melville Bay in less than 50 days.

Grose (1979) used data from this spill to validate a computer model that predicts slick thickness of spilled oil as a function of the physical characteristics of the oil and weathering by sheen formation and evaporation (wind speed dependent). This model will be considered in more detail in the modeling section (Chapter 5); however, it should be noted that the model was designed to account for three observed phenomena. The first phenomenon was that the oil did not form a single pool, but was composed of numerous patches of thick oil surrounded by thinner sheen. The second phenomenon was that the thickness of the patches was a function of the bulk physical properties of the oil and local environmental conditions, including wave height, water temperature, and wind speed. The third observed phenomenon was that weathering of individual components was dependent on the physical/chemical properties of the components in the original oil. Losses of oil from dispersion into the water column were ignored in the model because they were observed to be relatively small in the field (calm conditions persisted during the Potomac spill) and because of a lack of parameterizations for the pertinent process. Also, there were no data available on the probability size distribution of the patches as a function of either sea state or spill rate, although Grose did suggest that the mean size of patches will increase as the sea state decreases and the spill rate increases. Thus, for the model, he used an equal probability for all sizes of initial oil patch volume. The model did not include parameters for emulsification or incorporation of water, which would change specific densities, viscosity, and the total surface area and volume of the patches. However, as noted above, water-in-oil emulsion formation was not observed in the field during this spill. The best fit of the data to the computer predicted outcome occurred with a 0.01 m^2 patch and a wind speed of 2 m/sec. The largest discrepancies between predicted and observed behavior occurred for the size of the patches, which were reported to be an order of magnitude smaller in the field (0.018 cm^2) than the sizes predicted by the model (0.4 cm^2).

AMOCO CADIZ

On March 16, 1978 the supertanker Amoco Cadiz grounded off the coast of North Brittany, France, and its entire 220,000 ton cargo of Arabian crude oil and Bunker C fuel was released to the environment. Mousse containing 50-70% water formed almost immediately after release of oil from the vessel (or even prior to release because of mixing from wave and tide action in the bottom of the ruptured tanks). The oil and freshly released mousse consisted of 39% saturated hydrocarbons, 34% aromatics, 24% polar materials, and 3% residuals (Calder and Boehm, 1981). Downward mixing of the oil and mousse into the water column was caused by turbulence in the nearshore waters. Concentrations of dispersed mousse ranging from 200 to 1,000 μg/liter were observed near the entrance of Aber Wrac'h, and concentrations in excess of 500 μg/liter were observed in much of the waters throughout the estuary (Calder and Boehm, 1981). Additional evidence of dispersed oil or mousse in the water column can be inferred from Wolfe et al. (1981). Gas chromatograms of aromatic hydrocarbon fractions isolated from transplanted mussel samples (suspended in cages in areas of high contamination) showed high levels of dibenzothiophene, alkyl-substituted dibenzothiophenes, and three and four ring aromatics, including phenanthrene, anthracene, benzo-a-anthracene, and chrysene, plus their alkyl-substituted homologs. Gas chromatographic profiles from the organism extracts closely resembled those of background mousse samples. These data suggested that the mussels incorporated substantial quantities of either particulate oil or particulate mousse. Other studies have indicated that these higher molecular weight components were not present in true solution in the water column.

In reference mousse samples collected with a bucket from the immediate vicinity of the broken tanker, nC-11 was the most abundant n-alkane, although an homologous n-alkane series from nC-8 to nC-30 was noted. The pristane to phytane ratio was near unity, and both nC-17 and nC-18 were more than twice as abundant as the nearest isoprenoids. Numerous alkyl-substituted aromatics, ranging from tetramethylbenzene to methylphenanthrene, were identified, as were a series of alkyl-dibenzothiophenes. The most abundant aromatic compounds in the parent mousse were naphthalenes, methylnaphthalenes, dimethylnaphthalenes, C3-naphthalenes, fluorene, and phenanthrene. In a sample believed to be approximately eight hours old, the normal alkane composition had been altered by evaporative weathering, and molecules boiling below nC-15 showed detectable losses relative to n-tetracosane. The nC-17/pristane and nC-18/phytane ratios were not detectably altered, however, suggesting that microbial processes had not been active up to this time (Calder and Boehm, 1981). The aromatic components in this sample were also altered by evaporation and dissolution, with removal of the alkylbenzenes and methylnaphthalenes relative to dimethylnaphthalenes. Concentrations of phenanthrene and dibenzothiophene were not appreciably reduced in this sample relative to those in the reference mousse.

In other samples which may have been in the water column longer, and in numerous samples of buried or sedimented oil, microbial activity was particularly important to degradative weathering of the oil. In the dispersed oil-in-water samples collected near the entrance of Aber Wrac'h, a large reduction in nC-17/pristane and nC-18/phytane ratios and an enrichment of branched, cyclic, and aromatic hydrocarbons relative to the n-alkanes indicated that biodegradation was occurring at a faster rate than evaporation or dissolution. Aminot (1981) found that the lack of oxygen was one of the most critical environmental limitations to water column biodegradation of Amoco Cadiz oil. He related nutrient water chemistry to levels of persistent Amoco Cadiz oil; deficiencies of nitrogen, phosphorous, and oxygen were correlated with the presence of oil in the water column, and provided indirect evidence of in situ oil biodegradation. Ward (1981) stated that once the oil was deposited in the sediments, oxygen limitation again played an important role, in that both n-alkanes and aromatic compounds were more persistent in reducing sediments.

Following the wreck and initial dispersion of oil from the Amoco Cadiz (March 16-March 30, 1978), Hayes et al. (1979) undertook field studies to attempt a mass balance estimate of the total oil on the beach from March 19 to April 2, 1978 and again during the period of April 20 to April 28, 1978. During the first two weeks of the spill a total of 72 km of coast was heavily oiled. Using a value of 887 tons of oil/km of shoreline, Hayes at al. (1979) estimated that 64,000 metric tons of oil were deposited along the coastal zone, representing approximately one-third of the total amount of oil lost from the Amoco Cadiz. The remaining two-thirds were believed to be lost by evaporative processes, or represented by oil masses remaining on the water surface, sinking to the bottom, and/or mixing into the water column. One month later a total of 10,000 metric tons of oil could be accounted for on the beaches; this represented an 84% decrease in the oil mass found along the shore in the first visit. More coastline was covered at this time, however, due to wind shifts on April 2, 1978, which caused oiling of previously clean coastal areas south of the wreck site. One month after the spill, the impacted area had increased to 213 km of lightly oiled beaches and 107 km of heavily oiled beaches. Hayes et al. (1979) reported that the geomorphology in the coastal zone was very important to the distribution of the oil and mousse. In many areas, oil/mousse mixtures were found to settle in pools around boulders, bar troughs, and marsh pools; intertidal rocks, joints, and crevices were also covered with oil. Because of the wide variety of sediment, headland, and marsh types in the area affected by Amoco Cadiz oil, the authors developed an Oil Vulnerability Index correlating shoreline type with degree of oiling to estimate long-term impacts and fate. As in other spills, the exposed rocky coastlines were cleaned more rapidly than sheltered coasts. Tidal flats and estuary marsh systems retained oil for longer periods and were found to be extremely vulnerable.

Numerous protracted studies at several of these sites have been completed in the years following the spill. Results from a number of these studies were summarized by

Gundlach et al. (1983). The following paragraphs describe only the most salient features with regard to the fate of stranded mousse in different intertidal regimes. In general, the degree of oiling and persistence of stranded oil can be directly correlated with intertidal energy and substrate type.

Calder and Boehm (1981) reported that the mousse deposited on sandy tidal flats at Aber Wrac'h contained a lower percentage of saturated aliphatic hydrocarbons compared to the aromatic components, which presumably reflected microbial degradation of the saturated materials. During the first four months following the spill, both saturated and aromatic hydrocarbons were rapidly lost from surface sediments. However, after approximately five months the concentrations remained relatively constant for the next eight months at about 5% of their initial values. During the seven month period following the spill, the unresolved complex mixture, observed in gas chromatographic profiles of sediment samples, decreased from 410 ppm to 80 ppm, and the n-alkanes in the nC-10 to nC-35 range decreased from 35 ppm to 1.5 ppm. During the first month, the pristane plus phytane concentration dropped from 6.5 to 3 ppm (factor of two) and the nC-17 plus nC-18 concentrations dropped from 1.8 to 0.3 ppm (factor of six). This higher loss of the straight chain aliphatics compared to the isoprenoid compounds clearly demonstrate the importance of bacterial activity in the reasonably well oxygenated sediments during the first month. After seven months the isoprenoid compounds were dominant in chromatograms of the sediment extracts, and only a few n-alkanes were detected across the entire boiling point range of nC-10 to nC-34. Gravimetric analyses of aromatic fractions showed similar losses, with total aromatics dropping from 630 ppm in April to 90 ppm in late October. The total resolved component concentrations also dropped from 29 ppm to 2 ppm over the same time frame. In March 1979, one year after the spill, alkyl-substituted phenanthrenes and dibenzothiophenes were still easily detectable in the sediments above 20 cm. Below 20 cm depth in the sediments, aliphatic and aromatic hydrocarbons comprised materials from a biological origin.

Boehm et al. (1981) studied compositional changes in beached or sedimented oil/mousse at a variety of coastal environments, including rocky shores, sand flats, coastal embayments, tidal mud flats, and salt marshes. The authors found that much of the oil appeared to be relatively long-lived in the sediments, although some fractions were rapidly weathered and removed. The compounds most resistant to chemical and biological weathering were alkylated organosulfur compounds (dibenzothiophenes), alkylated phenanthrenes, and polycyclic aliphatics (e.g., pentacyclic triterpanes and hopanes). It should be noted that the actual concentrations of alkylated phenanthrenes and dibenzothiophenes did decrease, although they also became more predominant relative to the other aromatics compounds as weathering progressed. Boehm et al. (1981) also detected a number of pyrolitic polynuclear aromatic hydrocarbons (PNAs) in the sediment samples collected from the spill area and from control sites. These compounds included fluoranthenes, pyrene, benzofluoranthenes, benzopyrenes, and pyrelene at parts per million levels. Concentrations of these PNAs were fairly uniform with depth in the core, and the presence of these materials were attributed to background hydrocarbons, presumably from combustion sources.

At Portsall (station AMC-4) the surface sediment was primarily coarse sand. After an initial heavy mousse coating, absolute concentrations decreased rapidly, although the n-alkane to isoprenoid ratio remained relatively high until December 1978. The December 1978 aromatic data indicated that the remaining mousse/sediment mixture contained an abundance of alkylated two and three ring aromatics and dibenzothiophenes. Weathering reactions then appeared to increase between December 1978 and March 1979, presumably as a result of increased mixing and resuspension of the sediments accompanying winter storm activities. At Ile Grande salt marsh (station AMC-18), the degradation rate of deposited mousse appeared to be slower. At St. Michael-En-Greve (station AMC-19), a massive kill of benthic fauna was observed in the sand flat due to heavy oiling. Sediment samples collected in December 1978 and March 1979 showed aliphatic and aromatic profiles that were indicative of highly weathered oil. Several offshore stations adjacent to the Bay of Morlaix (stations AC-42, 138, 371, and 453) were also studied. In April 1978, immediately after the spill, the sediments from these areas contained only low quantities of _Amoco Cadiz_ oil; a mixture of chronic and background hydrocarbons and terrigenous runoff material, and perhaps small amounts of _Amoco Cadiz_ oil, were suggested by GC traces. However, in July 1978, weathered _Amoco Cadiz_ oil dominated the aliphatic and aromatic hydrocarbon assemblages, presumably due to extended offshore transport. By November 1978 overall concentrations were significantly reduced, and GC profiles illustrated that the oil had achieved a final weathering stage. Almost all of the resolved features, other than hopanes, were removed and not discernible above a large bimodal UCM. _Amoco Cadiz_ oil was weathered and removed most rapidly from the sand beach at St. Michael-En-Greve and from offshore sediments adjacent to the Bay of Morlaix. In contrast, oil levels remained high in intertidal mud flats of Aber Wrac'h and at Ile Grande salt marsh, and although the oil composition changed rapidly, the oil was found to degrade slower in the fine substrates than in coarser grained sediments. In general, buried oil appeared to be preserved over the 15 month period of the study. The deeper buried oils apparently escaped appreciable degradative weathering, presumably due to anoxic conditions at depth in the sediments. In general, it was found that nitrogen, sulfur, and oxygen compounds (NSO content) increased in the residues as hydrocarbon concentrations dropped. However, the degree of oiling was important in this context because at the more heavily oiled sites weathering appeared to be slower at both high and low energy shorelines. Thus, Boehm et al. (1981) concluded that the most important factors responsible for the observed relative weathering patterns were: extent of oiling > shoreline energy > sedimentation oxidation state >> distance from the wreck site.

Vandermeulen et al. (1981) examined the high energy rocky shoreline on Ti Saozin Island and a lower energy tidal mud flat at Aber Benoit for mousse content in April 1978, June 1978, and January 1979. In April 1978, the Ti Saozin shoreline was uniformly oiled and mousse was deposited along the upper beach at several stations. In June 1978, dried mousse completely covered one station while patches were found at other locations at an approximate one meter lower tidal level. Self-cleaning of the mousse at the high energy shoreline of Ti Saozin was nearly 90% complete within nine months after the spill. Other areas appeared to be visually clean of oil, although sheen was noted in some of the tide pools. By January 1979, only traces of mousse mixed with beached debris were found at one station, whereas some tar could be seen in rocks, cracks, and under boulders at another location. The rest of the slope appeared to be clean of the oil. In the tide pool samples from this location, however, only the water soluble components of the stranded oil were examined. Thus, the presence of highly insoluble asphaltenes and other residuals was not studied. These materials may have been present under rocks or sediments or bound-up in interstitial debris. Vandermeulen et al. (1981) did note that the rapid cleaning of the foreshore area of the intertidal zone (with traces of some oil remaining in the high tide zone) at Ti Saozin was faster than for similar shorelines in Chedabucto Bay after the *Arrow* spill. In the latter case, high energy shorelines were cleaned after approximately one year. Vandermeulen et al. (1981) stated that differences in oil viscosities and air and water temperatures at the time of the respective spills (February and April) may be responsible for the observed differences in residence or self-cleaning times at the two spill locations. The self-cleaning time for Ti Saozin was estimated at about two to four times faster than for similar shorelines in Chedabucto Bay.

In the lower energy tidal mud flats at Aber Benoit, large amounts of petroleum hydrocarbons were observed in the sediments of the inshore mud flats and along the vegetated fringe at the edge of the Aber Benoit River nine months after the spill. Synchronous excitation emission fluorescence scanning spectroscopy (SEES) spectra of the parent mousse and stranded tar collected in January 1979 showed that a number of two, three, four and five ring, and larger, aromatic compounds persisted at Aber Benoit where the tidal range was nine meters during high spring tides. Thus, in contrast to the Ti Saozin sites, the sediments at Aber Benoit remained heavily oiled as of 1979, with hydrocarbons from the *Amoco Cadiz* present at levels well above those found in control sediments. In the lower energy regimes it appears that the long term persistence of this oil may extent into tens of years.

IXTOC I BLOWOUT, BAY OF CAMPECHE, GULF OF MEXICO

Background

On June 3, 1979 the Petroleos Mexicanos (PEMEX) exploratory well, IXTOC-I, blew out in approximately 60 m of water at a site about 80 km northwest of Ciudad del Carmen in the Bay of Campeche, Gulf of Mexico. Within one week, the blowout had discharged more oil than the largest oil spill in U.S. history, and by late July IXTOC-I had become the largest oil spill in history, exceeding the 230,000 tons of oil that were lost from the *Amoco Cadiz*. The IXTOC-I blowout continued losing oil until successful drilling of a relief well and eventual capping on March 23, 1980. At that time, estimates of oil spilled over the ten months were in excess of 476,000 tons (Gundlach et al., 1981).

Extensive areas of the Gulf of Mexico and portions of its bordering beaches were eventually covered by oil released from the blowout (Figure 7). After the first three days, slicks 93 km long and several kilometers wide were reported (OSIR, 1980). Nine days after the blowout, the main slick was 180 km long and up to 80 km wide, and was moving in a westerly direction with a velocity of about 0.5 knots.

NOAA trajectory models indicated that much of the spilled oil reaching a branching point off Tampico would turn south and likely impact Mexican coastlines. On July 20, oil washed ashore at Coatzacoalcos, and by July 24 oil was washing ashore north of Tampico, on the beach at Cabo Rojo near Tuxpan de Rodriguez, and at Veracruz. By July 29, sheen and mousse were about 180 km south of Brownsville, Texas, and large patches were observed from Tampico to a point 230 km south of Brownsville. These slicks/patches reportedly were moving north at a rate of about 0.5 knots, and mousse aggregates 2 to 3 m in diameter and several centimeters thick formed a broken line more than 16 km long (OSIR, 1980).

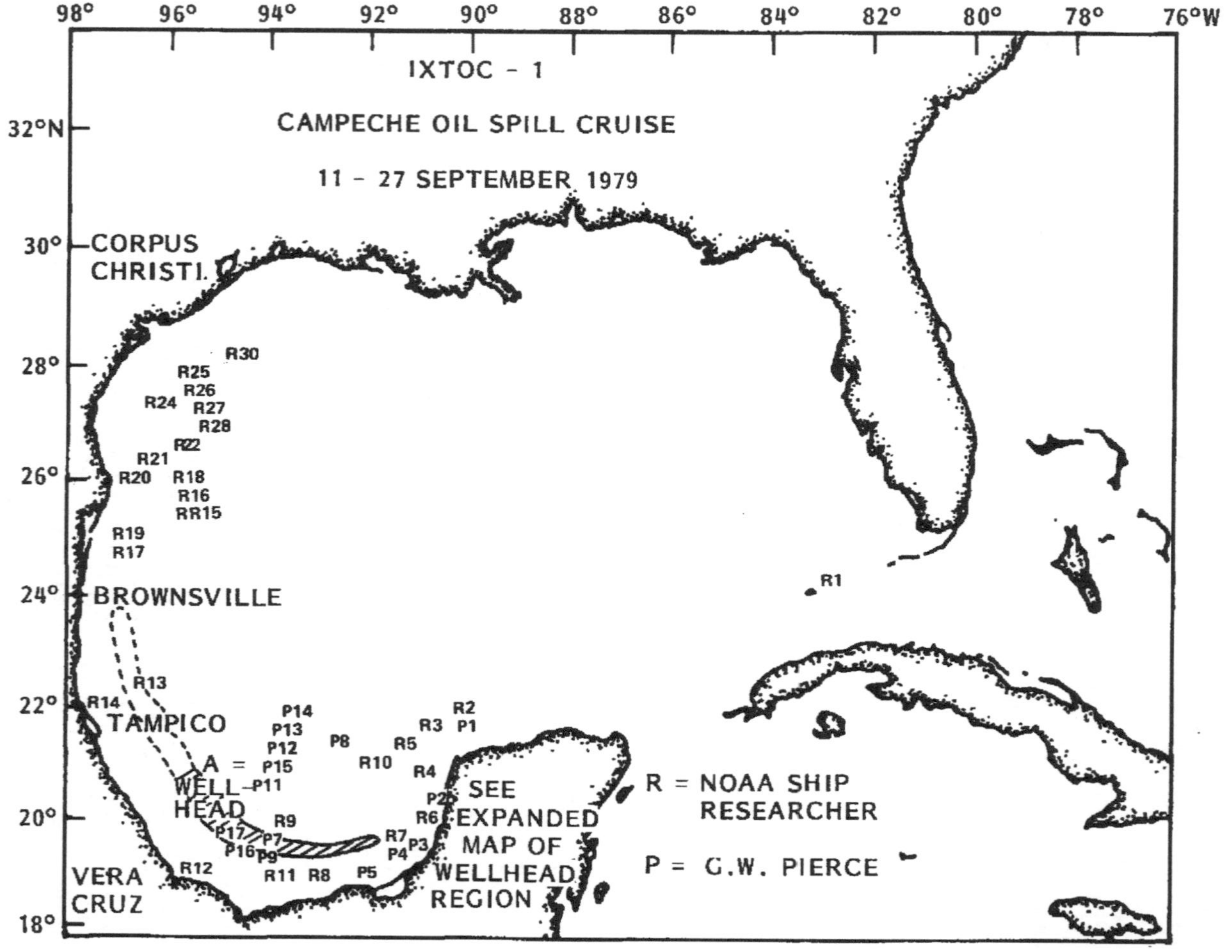

Figure 7. NOAA SHIP RESEARCHER AND G.W. PIERCE STATION LOCATIONS AND APPROXIMATE TRACK OF OIL SLICK FROM IXTOC I BLOWOUT.

Patches of IXTOC-I mousse were first observed in U.S. waters about 70 km off of Brownsville, Texas on August 6. A week later, large concentrations of sheen were observed between Port Mansfield and Brownsville, about 48 km offshore, and ribbons of mousse up to 16 km long and 85 m wide were observed 32 km offshore. By August 13 patches of mousse and sheen were up to 530 km out to sea between Tampico and Tuxpan de Rodriguez Cano, and an 11 km patch of heavy mousse concentrations was reported near Tampico. Dense oil coatings were reported on the beaches of Tuxpan de Rodriguez Cano, near Veracruz, and at Laguna Madre. Heavy concentrations of mousse washed ashore along 80 km of Texas coast with the return of southeasterly winds on August 18, 1979.

Oil finally stopped washing onto Texas shores in mid-September when the seasonal change of winds and currents offshore began reversing the northward flow. Estimates in late August suggested that oiling of beaches had reached a maximum when up to 3,900 tons of oil covered Padre Island and Port Aransas, Texas, beaches (Gundlach et al., 1981), although Roy Hahn (TAMU) and John Robinson (NOAA) estimated that a total of 10,000 tons of oil had washed onto Texas shores (OSIR, 1980). Researchers and cleanup workers noted that in September 1979, IXTOC mousse had collected in the sand beaches of Padre Island to depths up to 30 cm.

Large quantities of IXTOC mousse and oil also impacted many areas of the Mexican coastline; however, details on the extent of oiling and cleanup operations are not available. The Oil Spill Intelligence Report (1980) stated that approximately 2,100 gallons of oil were recovered from Tuxpan de Rodriguez Cano. At Laguna de Terminos, Ciudad del Carmen, oil was unable to cross the natural barrier at the interface between the saltwater and fresh water of the lagoon; thus, only minor quantities of oil entered the sensitive biological habitat during the first few months of the spill. Increased rains and heavy river flow extended the natural offshore barrier, thus protecting the lagoon and associated shrimp nurseries.

In late September with a change in currents at the wellhead, oil slicks in the Bay of Campeche were moving south at about 2.7 knots. Large amounts of oil were sited in October between the IXTOC-I well and the northwestern tip of the Yucatan Peninsula, and scattered sheen was noted up to 560 km northeast of the well. Heavy coverings of IXTOC mousse (between 15 to 30 cm deep) were reported in November, 1979 at several locations on the western shore of the Yucatan Peninsula.

There was much concern about IXTOC mousse and its eventual impact on U.S. and Mexican shores. In response to this concern, the United States EPA and the National Oceanic and Atmospheric Administration (NOAA), among others, studied the weathering and fate of the released oil. Perhaps the most extensive study of weathering effects on IXTOC crude at sea occurred during NOAA-sponsored cruises at the wellhead and along the eastern coasts of Mexico and Texas in September, 1979 (Preliminary Results from the September 1979 Researcher/Pierce IXTOC-I Cruise, Symposium Proceedings - 1980).

A Canadian Environmental Protection Service study in June 1979 (Ross et al., 1980) reported that the IXTOC-I oil immediately formed a viscous water-in-oil emulsion. However, the Researcher/Pierce studies in September demonstrated that mousse was not forming immediately after release of the oil at the wellhead, but was the product of extensive evaporation, dissolution, and, presumably, photochemical weathering. Based on chemical analyses, the emulsion at the water surface contained about 70% water with a viscosity of 350 cP and a specific gravity of 0.99. The stable water-in-oil emulsion contained very fine droplets of water coated with thin viscous layers of oil, and formed a 1 to 3 mm slick on the water surface. In contrast, the oil collected during June was almost devoid of the lighter fractions (boiling points less than nC-10) that were burned, evaporated, or dissolved prior to water-in-oil mousse formation. One possible explanation for the differences between the June and September studies is that the composition of the oil may have changed. Specifically, as the upper levels of the IXTOC-I oil reservoir were depleted, the later-released oil could have changed chemically during migration through sediments within the formation. Another hypothesis was that with decreased flow rates in August (claimed by PEMEX) the accompanying decrease in mixing energy in the water column might have affected mousse formation.

The original Researcher/Pierce cruise plan called for collecting oil and mousse samples and studying the fate and chemical alterations along the semi-continuous slick which had been reported along extensive portions of the western Gulf of Mexico. Immediately prior to sampling, however, Hurricane Henri passed to the north of the Bay of Campeche generating winds in excess of 24 knots and six meter seas in the vicinity of the wellhead. The storm caused much of the semi-continuous slick to be dispersed into the water column. After the hurricane, large patches of oil or mousse were not observed beyond 40 to 50 km from the well. Thus, subsequent studies of the weathering, mousse formation, and subsurface oil movement during the cruise were more intensive near the wellhead, but relatively less extensive at greater distances from the source of the oil. Nevertheless, much of the observed oil weathering behavior at the wellhead was affected by the intense storm activity.

Extensive resuspension of the bottom sediments occurred in the Bay of Campeche following passage of Hurricane Henri; suspended particulate loads were estimated as high as 100 mg/l. This highly turbid water was observed along much of the coastline of the Bay of Campeche area, and the distinct turbid/clear-water boundary appeared to be associated with oceanic fronts. At the time of the Researcher/Pierce study, the area immediately around the IXTOC-I wellhead was within the turbid waters (Figure 8).

In related mineralogy studies, Nelson (1980) reported that no mineral species were present in the suspended particulate material (SPM) that did not also occur in the bottom sediment. Thus, although large amounts of river-derived SPM were introduced following the intense rainfall associated with the storms, most of the turbidity in the vicinity of the wellhead was the result of resuspended sediments. Six terriginous and four carbonate minerals were identified in the SPM. The terriginous suite contained quartz, pelagic feldspar, and the clay minerals chlorite, smectite (montmorillonite), illite, and kaolinite, whereas the carbonate suite contained arragonite, calcite, dolomite, and magnesium calcite. Many of these minerals actively adsorb particulate and dissolved hydrocarbons from the water column (see Jordan and Payne, 1980), and their presence probably affected oil behavior.

Observations at the Wellhead

During the Researcher/Pierce cruise in September 1979, the slick at the wellhead was flowing in a northeasterly direction. At times the plume took sharp meanders, generally to the south. However, the direction of the oil flow on the surface was not correlated directly with wind speed or direction. Because the oil was released at depth, water column currents in the vicinity of the well controlled the initial direction of oil flow, and on several occasions the wind direction was 180° counter to the direction of plume movement.

Immediately over the wellhead, a large fire about 50 m in diameter and seven meters in height was burning on the water surface. It appeared from the yellowish-orange color and general absence of smoke that the fire was consuming primarily lower molecular weight gases and liquid hydrocarbons. Much of the oil released from the wellhead was not exposed to the fire as it was advected or removed from the source by sub-surface currents. There appeared to be continuous upwelling of water within and surrounding the fire, and within 20 m of the burning zone smaller random surface flares appeared periodically, but were later extinguished by the surrounding turbulent surface waters. Tan to rust-colored oil was observed on the water surface around the fire and extending about 150 m on each side of the flames. This oil, in conjunction with the rising droplets released from the well, then formed a semi-continuous slick which extended towards the northeast. The oil slick definitely was not

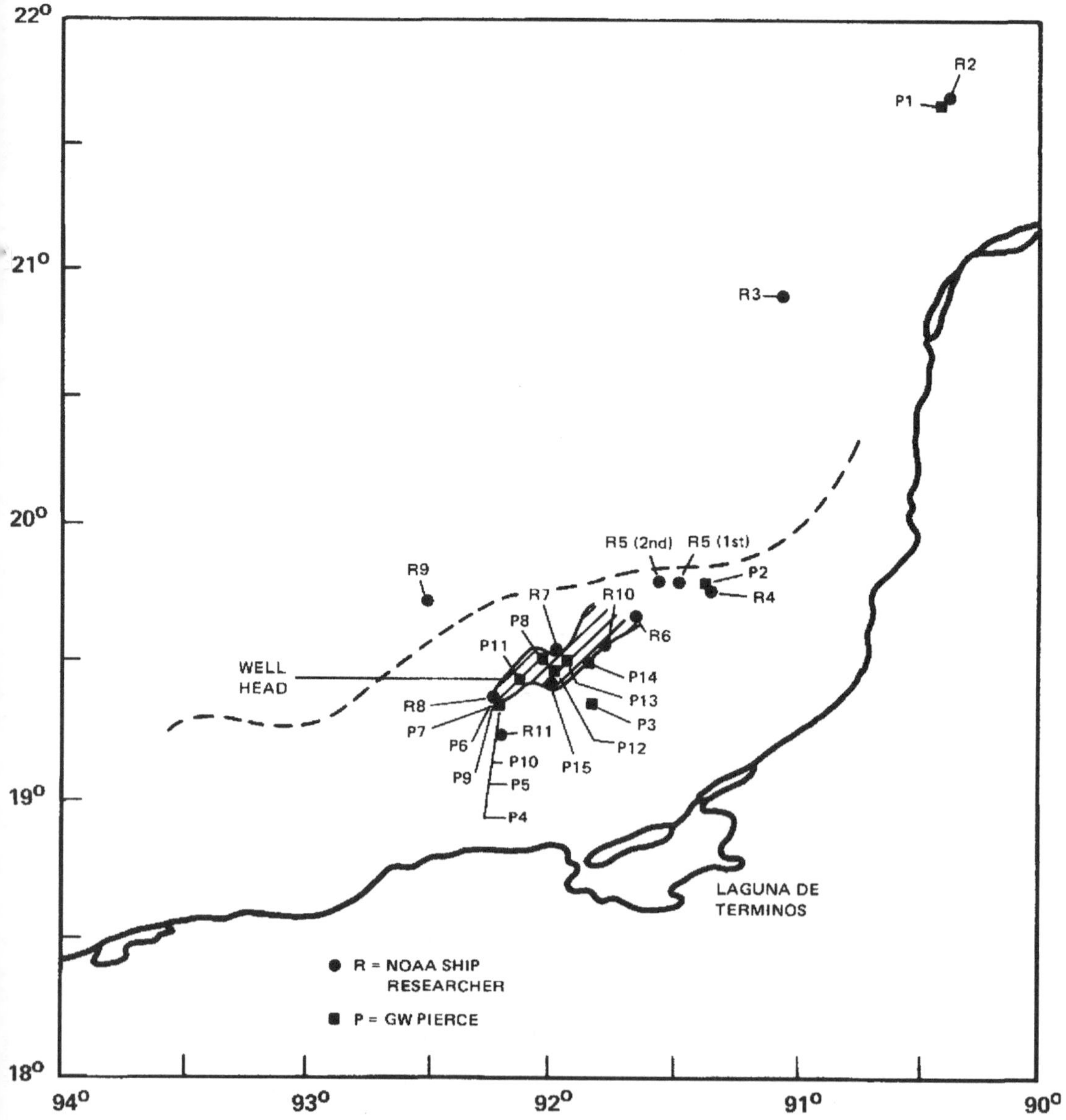

Figure 8. IXTOC-I CAMPECHE OIL SPILL CRUISE, 11-27 SEPTEMBER 1979. EXPANDED WELLHEAD REGION.

– – – – GREEN (SEDIMENT LADEN) BLUE WATER FRONT
——— APPROXIMATE OIL PLUME TRACK

emulsified into a stable mousse, although it was a light brown color and probably contained some water as reported earlier by Ross et al. (1980). Immediate formation of a stable water-in-oil emulsion or mousse was not observed. At distances of 3 to 50 nautical miles from the wellhead, large patches of stable water-in-oil emulsions were observed with varying sizes and shapes. However, it was often difficult to distinguish freshly generated mousse from more weathered materials which presumably had been formed at an earlier period and then re-entrained into the slick by eddy currents near the wellhead.

In the vicinity of the wellhead the physical state of the oil was uniquely different within each of five distinct zones. The relative size and position of the zones appeared to be a function of many factors including sunlight intensity, wind stress, and flow rate (Atwood et al., 1980).

The first zone was characterized by the continuous low viscosity, light brown, water-in-oil emulsion on the surface. This material existed in the immediate vicinity of the flames and extended for no more than a few hundred meters down plume. This zone represented 100% coverage of the water surface with an oil film ranging in thickness from 1 to 4 mm.

The second zone was characterized by a 30 to 50% coverage of the sea surface by a light brown water-in-oil emulsion in randomly oriented streaks. This area started a few hundred meters down plume from the burn and extended out to a maximum of one to two nautical miles, depending on wind stress. At times, however, this zone was not distinguishable and its presence appeared to depend very critically on environmental conditions.

The third zone was characterized by a 20 to 50% coverage of the sea surface by light brown water-in-oil emulsions oriented in Langmuir "streaks" surrounded by a light to heavy sheen of oil. These streaks were parallel to the wind direction and varied in width from a few centimeters to several meters and in length from one to tens of meters. These dimensions varied with wind stress. This zone also extended from as close as a few hundred meters from the flames to several nautical miles down plume.

The fourth zone was characterized by a darkening of the light brown water-in-oil mixture until the streaks were black. This was assumed to result from oxidation of the oil, and the oxidation rate appeared to be dependent on sunlight intensity. A slight color gradation was noted, with many of these streaks blackened in the center and light brown at the outer edges. At times these "streaks" coalesced into long lines of blackened oil that extended for several kilometers. Along the edges of these "streaks" small balls of chocolate mousse-like emulsion were observed to break off but, as will be discussed later, readily coalesced into larger masses upon contact. At other times the wind rolled portions of the slick up onto itself, which was observed as a mechanism for mousse agglomeration. In some instances mousse balls reached 20-40 cm in diameter and/or coalesced into huge "rafts" of mousse up to 50 or 60 meters in diameter. Several rafts of this type were up to one meter thick. This zone began 5 to 15 nautical miles from the wellhead and extended out to about 20 nautical miles. Detailed observations of formation of this type of material on a micro-scale were made during a 26 mile down plume transect on the GW Pierce (these observations will be discussed in detail later).

The fifth zone was characterized by an extensive light to heavy sheen of oil that covered >50% of the surface, usually in the form of Langmuir lines. This zone overlapped Zones 2, 3, and 4 and extended out to the furthest extremity of the plume.

As noted in the introduction, these descriptions may have been relevant only for the period of the Researcher/Pierce cruise, and they may have been influenced by the intense storm activity from Hurricane Henri. At several stations surrounding the wellhead no distinct boundary between these various zones was apparent as the zonal characteristics often tended to interweave at the transitions.

At several stations, relatively large (20 cm) sized patches of mousse appeared to have had been stained with fresh oil. Sheen was observed coming from some mousse patches but not from others. To a limited degree, the presence or absence of sheen emanating from the external surface of the mousse (or the interior portions if the samples were broken and re-introduced into oil-free water) appeared to correlate with stickiness. The lack of this sheen or "stickiness" was taken as evidence that such samples may have been more weathered before formation. The sizes and dimensions of the mousse patches varied, ranging from baseball (10 cm) sized balls to elipsoidal 20 to 30 cm diameter patches ranging from 0.5 to 1 cm thick. Elsewhere, solid 1 to 2 m^2 patches were observed with thicknesses ranging from 5 to 10 cm. When 20 to 30 m diameter by 0.5 m thick "rafts" of mousse were encountered, discolored (yellow) water was often noted floating on top of the mousse (particularly in calm sea conditions). Most of the mousse balls or pancakes were a light chocolate brown in color and had a brain-like or "popcorn ball" appearance; many of these appeared to contain large (several cm) pockets of water. Several such mousse balls were observed to melt and separate into water and oil when they were removed from the water column and exposed to sunlight. Following separation the color changed from a light chocolate brown to a dark brown/black or mahogany. Attempts to separate such mixtures by heating and centrifugation onboard the Researcher were unsuccessful. At other stations, mousse was associated with large quantities of marsh detritus and sugar cane stock. It was hypothesized that these materials provided sites of nucleation upon which additional aggregation of mousse could occur.

Subsurface Transport and Weathering of IXTOC Oil and Mousse

Large amounts of the particulate oil droplets initially released at a depth 60 m were transported horizontally for distances of 20 to 30 km before the oil surfaced and was exposed to evaporative weathering. During the spill investigation acoustic profiling was performed both by Walter and Proni (1980) and by Macaulay et al. (1980), and subsurface water and oil-in-water dispersions immediately beneath the surface slick were collected and studied by Boehm and Fiest (1980a, b). Concentrations of particulate and/or colloidal petroleum in the top 20 m of the water column within 25 km of the wellhead (i.e., beneath the surface slick) exceeded maximum concentrations observed in the other spills such as the Ekofisk Bravo blowout (300 μg/l; Grahl-Nielsen, 1978); Amoco Cadiz spill (350 μg/l; Calder et al. 1978); and the Argo Merchant spill (450 μg/l; Gross and Mattson, 1977). The total amount of petroleum in the top 20 m of water within 25 km of the well at the time of the sampling cruise was estimated at 20,000 gallons (70 million grams) or approximately 1% of the total oil mass observed on the surface. Boehm and Fiest (1980a) hypothesized that for this type of sub-surface blowout, about 1% of the total higher molecular weight hydrocarbons released would be present in the water column for considerable distances from the wellhead. Approximately 90% of this material was believed to be associated with the particulate fraction and about 10% present in the truly dissolved state.

Before passage of Hurricane Henri, the water column was vertically stratified, and the transport of much of this dispersed oil appeared to be associated with density gradients in the water column. Boehm and Fiest reported subsurface particulate material at distances up to 25 to 30 nautical miles from the wellhead. Payne et al. (1980a) observed detectable levels of dissolved and particulate hydrocarbons above, but not below, vertical density gradients at distances up to 20 nautical miles from the wellhead. After the passage of Hurricane Henri, when the water column in the vicinity of the wellhead was vertically mixed from the surface to the bottom, higher levels of particulate hydrocarbons were noted in samples collected five meters above the bottom (Payne et al., 1980a). It was not possible, however, to differentiate the oil associated with particles settling out of the water column from oil associated with resuspended bottom sediments. Following the storm it was noted that concentrations of dispersed or SPM bound oil were generally one to two orders of magnitude greater than concentrations of oil in the truly dissolved state (Payne et al., 1980a). The particulate phase contained primarily the higher molecular weight aromatics, whereas the dissolved phases showed evidence of preferential dissolution of the lower molecular weight aromatics and alkyl-substituted naphthalenes. Analysis of whole water samples collected beneath the slick showed the presence of both low and high molecular weight compounds associated with the dispersed whole oil droplets (Boehm and Fiest, 1980a).

By the time most of the subsurface oil droplets had reached the surface, considerable weathering and removal of lower molecular weight components had already occurred. Thus, many aromatics were lost from dissolution rather than evaporation because of the sub-surface release. Payne et al. (1980b) measured quantities (up to 100 μg/l) of benzene, toluene, ortho-, meta-, and para-xylenes and other selected lower molecular weight aliphatic and aromatic compounds in the water column immediately beneath and adjacent to the surface slick. Air samples obtained on TENAX® traps immediately above the slick illustrated that dissolution rather than evaporation accounted for removal of many of the more water soluble components during transport through the water column. Only traces of benzene were observed in the air (particularly in comparison with toluene, xylene, and aliphatic and cyclo-alkane compounds with similar molecular weights and vapor pressures but with relatively lower water solubilities) at distances as close as 0.75 miles from the wellhead. Benzene was not detected by selected ion monitoring GC/MS (level of detection - 1 ppb) in the air samples collected above the slick, even though pre-field studies had demonstrated the applicability of the technique for quantifying benzene and alkyl-substituted lower molecular weight aromatic and aliphatic compounds. Brooks et al. (1980) quantified levels of benzene and other lower molecular weight aromatics in the water column. Their analyses also detected traces of these components in waters beneath the slick as close as 0.5 mile from the wellhead, although they were present at reduced quantities compared to levels of the higher molecular weight compounds above nC-10. These lower molecular weight aromatics typically constitute a fairly large portion of most crude oils, therefore, their rapid dissolution during transport from the bottom to the surface represented an important mechanism for partitioning oil components from whole oil droplets into the water column. After the oil surfaced, evaporative processes constituted the primary weathering mechanism. By the time the surface oil had traveled several miles from the slick (where incipient mousse formation was observed), most compounds with boiling points less than that of nC-11 had been removed (Payne et al., 1980b; Brooks et al., 1980).

Thus, while very high concentrations of oil were measured in the water column in the immediate vicinity of the wellhead (10,600 $\mu g/l$ within several hundred meters of the blowout to 5 $\mu g/l$ at a distance of 80 km), this oil immediately beneath the slick was generally in the form of a sub-surface plume of oil droplets suspended in the top 20 m of the water column (Fiest and Boehm, 1980). Only the more water soluble, mono-cyclic aromatics appeared to be in true solution in this region. These water soluble aromatics resulted from enhanced dissolution due to the high surface to volume ratio of sub-surface oil droplets.

At distances less than 10 nautical miles from the wellhead and away from the surface plume, water column concentrations of oil were very low and approached levels observed in more pristine environments (Payne et al., 1980a, 1978). As noted above, the majority of the oil observed at these outer stations was associated with suspended particulate material. It was hypothesized that the high concentrations of SPM acted to scavenge the remaining dispersed and semi-soluble components from the water column. Particle scavenging of subsurface oil could eventually lead to oil removal via sedimentation processes (Payne et al., 1980a). Interestingly, substantial quantities of IXTOC oil were not observed in sediment samples obtained near the wellhead; Boehm et al. (1980b) concluded that only 1% to perhaps 3% of the spilled oil mass partitioned into the sediments. These observations may be confounded, however, by the fact that the sediment samples were obtained after the passage of Hurricane Henri, and much of the recently deposited sediment load was resuspended and thus not easily sampled. Additionally, all of the sediment samples were obtained with a Smith-McIntyre grab sampler, which frequently disturbs or mixes the surface 1-2 cm sediment layer (Callahan and Soutar, 1976). Artificially low values of hydrocarbons measured in the sediments surrounding the wellhead could have resulted from either storm-induced resuspension of surficial sediments or loss or mixing of surficial sediments during sampling.

Personal Observations of Micro-Scale Mousse Agglomeration

A 26-mile downplume sampling transect was occupied by the vessel G. W. Pierce to monitor real-time chemical and physical changes in the surface slick as water-in-oil emulsification occurred. Continuous observations of the surface oil behavior, the loss of volatile compounds due to evaporation, and the eventual breakup of the continuous oil slick into 0.5 to 1 cm size flakes or droplets were made. Stable mousse formation was not observed up to a distance of 12 to 18 km from the wellhead. The age of the oil droplets or flakes at this point was estimated from 13 to 19 hours.

At that point (and time) an interesting agglomeration phenomenon of these small (1 to 2 cm), heavily weathered, oil flakes was observed in a narrow band along the leeward side of the vessel. With the main engines off for a 2 to 3 hour period on the evening of 19 September 1979, the G. W. Pierce was oriented perpendicular to the wind direction. Under these conditions, the combined effects of small waves generated from the rolling vessel and the wind (0 to 5 knots), which swirled over the lee side, caused the oil flakes to concentrate as a one meter wide band approximately 1 to 2 meters from the hull. This band was observed along the entire leeward side of the ship, whereas no such accumulation of flakes or tar lumps occurred along the windward side of the vessel. The well-weathered flakes were agitated by the waves from the gentle rolling of the ship and repeatedly contacted one another; consequently a number of smaller flakes were observed to aggregate into larger balls. Considerable silver sheen was also present within a concentrated band, although it appeared to spread further from the hull than did the flakes and the tar/mousse balls. Chips of styrofoam and pieces of paper were added to the flakes as markers to ensure that time series observations of these flakes would not be confounded by drifting into different oil/particulate mixtures. These markers remained with the oil droplets for 10 to 25 minutes before sinking and did not move more than 20 to 30 feet aft during observations. Within 45 minutes, the 0.5 cm flake sized droplets consolidated into 1 to 2 cm sized balls.

To determine if continued gentle agitation of these flakes would result in agglomeration of larger tar/mousse balls, a bucket cast of these flakes and tar balls was obtained. One 3 cm sized mousse ball and 10 to 15 small (0.5 to 1.5 cm) pieces were isolated. The tar ball and flakes were left on the water surface in the bucket, and the side of the bucket was tapped lightly, producing small (1 to 2 cm) standing waves. After five minutes, the smaller tar pieces began to aggregate when they bumped, and most of the small flakes eventually adhered to the larger piece. After 5 to 7 minutes, there was only one large mousse ball and two 1 to 2 cm sized flakes; the remaining flakes had been observed to stick to the ball. The tar ball grew in size by approximately 50% with a large irregular shape. From these observations it became apparent that a similar mechanism could produce the brain-like appearance of tar balls and pancakes observed at various locations around the wellhead. That is, as one layer of flake adheres to the next, a very irregular popcorn ball type matrix is generated which entraps appreciable quantities of water. Then, with continued wave agitation this agglomerate would be compressed into the irregular shapes which were characteristic of the tar balls observed during the cruise. Thus, this formation mechanism could also explain the appearance of the 20 to 30 cm size lumps located near the wellhead. Chemical characterization of these flakes later showed the composition of this material to be identical to that obtained from several of the larger mousse patches and from a large beached mousse sample obtained at Laguna Madre (Payne et al., 1980a; Overton et al., 1980b).

One hypothesis for mousse formation in the IXTOC spill was that the breakdown of the continuous oil slick proceeds through the dissolution and evaporation of lower molecular weight components, leading to a thin semi-continuous oil film. Photochemical oxidation and continued evaporation then caused this slick to break up into streamers or partially submerged droplets or flakes. Sheen formation continues to enhance the removal of many of the less viscous, lower molecular weight materials and promotes dissolution and evaporation. As the volatile and lower molecular weight components were removed gradually, the residual surface droplets (which resembled corn flakes or snowflakes) no longer had a runny consistency and appeared to be very sticky. At this stage, surface spreading is not important and the flakes exist as discrete particles. These tar flakes were observed from helicopters to collect in wind rows, along oceanic fronts and against the sides of vessels. Under the influence of turbulence and wave energy they gradually aggregate into larger particles within these convergence zones. With continued addition of small flakes, or after nucleation on drifting detrital materials such as sugar cane stock, these aggregates then form larger oil/mousse bodies. This hypothesis suggests that, at least in the case of this spill, mousse balls and tar balls were growth products from smaller particles, as well as breakdown products from the larger emulsified slicks. Such formation could occur in wind rows of oil flakes (which were observed) or in standing wave situations. Cleanup efforts in future spills might be enhanced by concentrating on such zones of convergence.

In this instance, the formation or aggregation of mousse appeared to occur only after removal of lower molecular weight compounds by evaporation and dissolution. Also, photo-oxidation of the slick appeared to enhance mousse formation, and it is believed that this process generated the appropriate surface active materials to stabilize the water-in-oil emulsion and mousse agglomerations. Overton et al. (1980a) identified a number of fatty acid methyl esters, normal fatty acids from C-7 to C-11, and branched fatty acids from C-9 to C-12 after photo-oxidation of fresh IXTOC crude oil. Alkyl-phenols, alylic benzoic acids with from one to six substituted carbon atoms, alkyl-substituted naphthols, alkyl-substituted naphthanoic acids, and alkyl-substituted phenanthroic acids were also identified, as well as alkyl benzothiophene acids and alkyldibenzothiophene acids. All of these compounds could act to stabilize water-in-oil emulsification or mousse formation.

Although many of these compounds would be expected to leach from the oil surface as they are formed (or during extended weathering), they might have promoted stabilization of the aggregation products of the smaller flakes. Evidence for this stabilization mechanism is suggested by the results from analyses of beached tar samples from Laguna Madre obtained by Overton et al. (1980b). The outer surface of this beached mousse consisted primarily of n-alkanes with limited numbers of aromatics, but no sulfur containing compounds. The middle portions of the mousse had higher aromatic contents and a number of sulfur-containing compounds. The weathering process thus appeared to remove oxygenated products from the surface but not from the interior of the mousse ball. Figure 9 shows the relative concentrations of selected polynuclear aromatics (PNAs) in the IXTOC crude oil and in the mousse flakes obtained at station PIX-13 (during observations of micro-scale mousse ball formation). Figure 10 shows the relative aliphatic distributions for the IXTOC crude, the mousse flakes described above, and the beached mousse sample collected at Laguna Madre (Payne et al., 1980a). Clearly, the alkyl-substituted naphthalenes, benzothiophenes, dibenzothiophenes, phenanthrenes, fluorenes, and pyrene are present at much higher relative concentrations in the mousse than the alkyl-substituted naphthalenes in the starting crude. Likewise, the distribution of the aliphatics shows that in the flakes and beached mousse, most n-alkane compounds with molecular weights less than nC-14 have been removed. Based on our observations in the field, we believe that this weathering process occurred before mousse flake or tar ball agglomeration/formation occurred. Table 6 presents ratios of nC-25 to nC-16, nC-25 to nC-19, and phytane to nC-18 with depth in the beached mousse collected at Laguna Madre. These data clearly indicate the absence of microbial degradation or differential weathering after mousse formation occurred. Again, it appears that overall chemical weathering occurred in the bulk of the oil prior to agglomeration into the larger 1 m^2 patch, which was eventually stranded in the upper intertidal zone several hundred miles from the wellhead.

Buckley et al. (1980) studied microbial processes associated to IXTOC mousse formation. They reported that Vibrio bacteria populations were associated with the mousse, although subsequent microcosm experiments showed little chemical evidence of microbial degradation of the bulk oil. The absence of microbial degradation was attributed to either the large influx of oil which masked any minimal changes due to microbial activity or, as Atlas et al. (1980) postulated, to the lack of nutrients in the water column in the Gulf of Mexico at the time of our observations. Another possibility was that microbial populations were inhibited by toxic

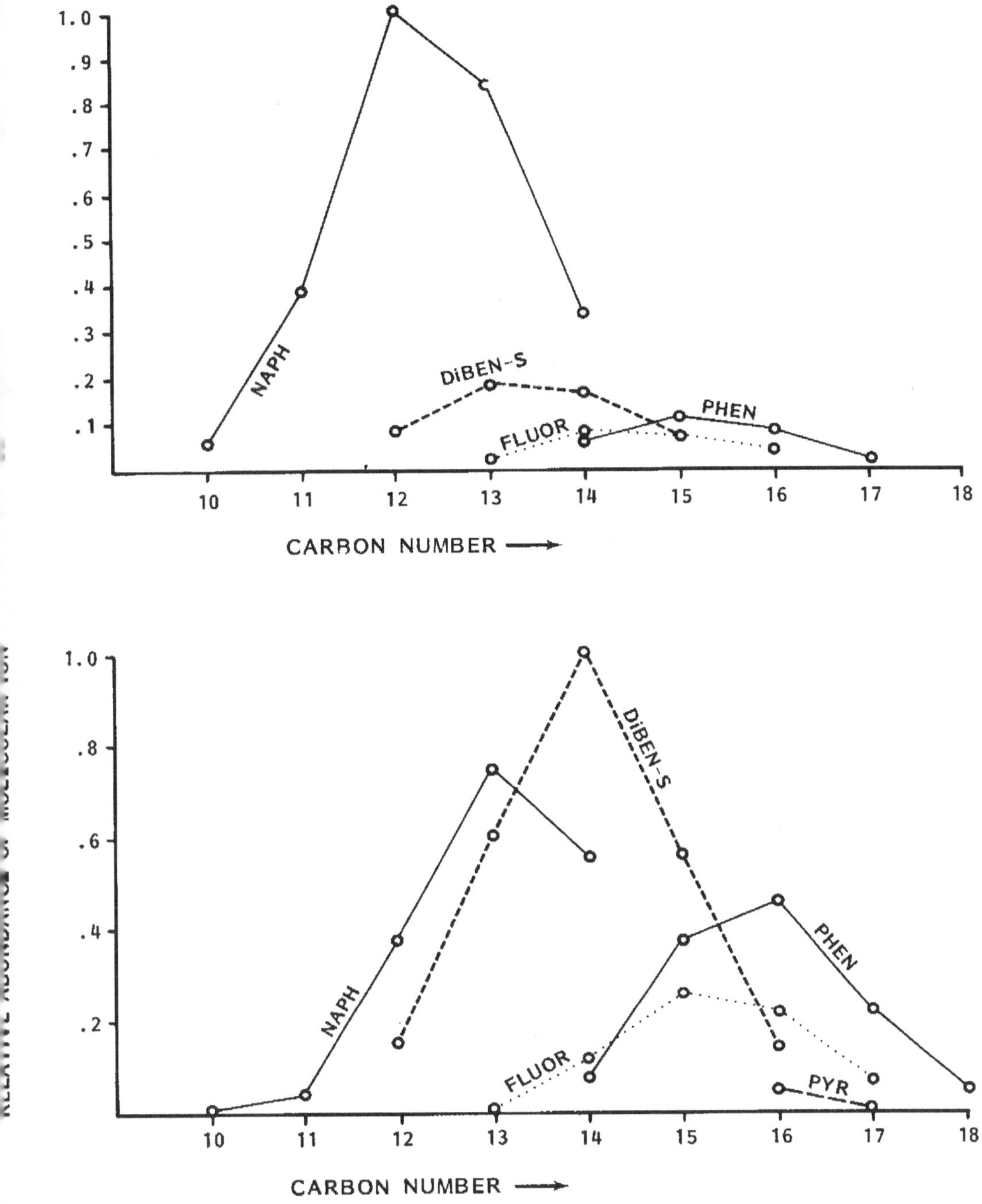

Figure 9. RELATIVE ABUNDANCE OF ALKYL-SUBSTITUTED POLYNUCLEAR AROMATIC HYDROCARBONS IN IXTOC CRUDE OIL COLLECTED ½ MILE FROM THE WELLHEAD (A.) AND FROM MOUSSE FLAKES COLLECTED 16-18 MILES FROM WELLHEAD DURING THE G. W. PIERCE DOWN-PLUME TRANSECT. NAPH = NAPHTHALENE; DIBEN-S = DIBENZOTHIOPHENE; FLUOR = FLUORANTHENE; PHEN = PHENANTHRENE; PYR = PYRENE. (FROM PAYNE et al., 1980a)

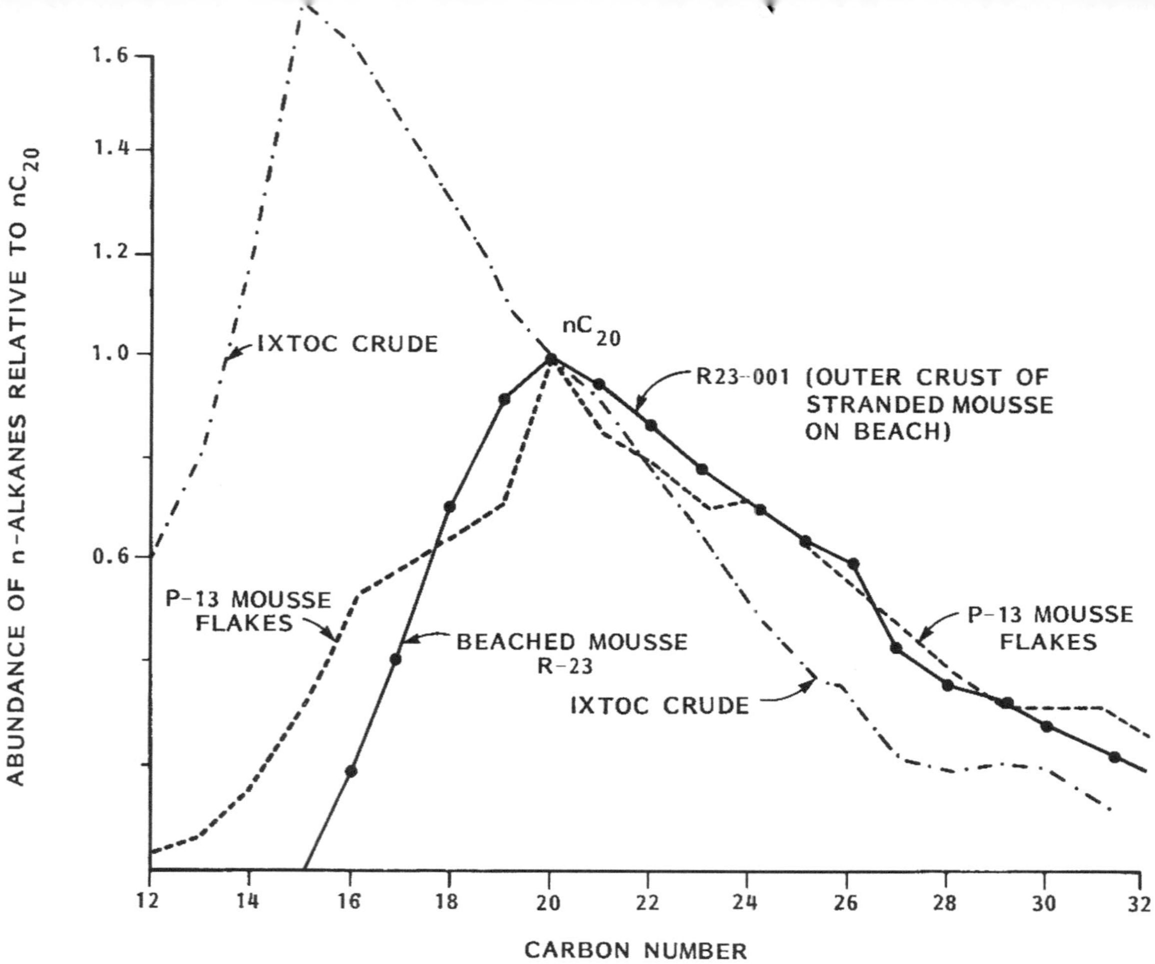

Figure 10. ABUNDANCE OF n-ALKANES (RELATIVE TO nC-20) IN BEACHED MOUSSE FROM LAGUNA MADRE (RIX-23), MOUSSE FLAKES COLLECTED 16-18 MILES FROM THE WELL HEAD (P-13) DURING THE G W PIERCE DOWN-PLUME TRANSECT, AND IXTOC CRUDE COLLECTED 1/2 MILE FROM THE WELL HEAD. (FROM PAYNE et al., 1980a.)

Table 6. Selected Component Ratios for the Dissected Mousse Sample Beached at Laguna Madre (Station RIX 23). (From Payne et al., 1980a.)

Sample *	Depth Below Skin	nC_{25}/nC_{16}	nC_{25}/nC_{19}	Phytane/nC_{18}
01	surface 3 mm	3.1	0.66	0.46
02	3-8 mm	4.3	1.05	0.42
03	8-13 mm	2.6	0.63	0.46
04	16-23 mm	2.3	0.61	0.44
05		--	Weeds	--
06	5-10 mm from bottom	2.4	0.69	0.44
07	bottom 5 mm	2.9	0.79	0.46

* 01 represents the sample from the air/mousse interface and 07 that from the sand/mousse interface.

effects inherent to the IXTOC oil (Buckley et al., 1980; Ron Atlas, personal communication). Specifically, in microcosm experiments, fresh oil-inhibited amino acid uptake and microbial activity could only be stimulated after five days. In contrast, mousse immediately enhanced amino acid uptake, which suggested that the removal of toxic components from the oil prior to mousse formation allowed the bacteria to subsist (if they were capable of hydrocarbon utilization). Overton et al. (1980b) analyzed samples from the fresh oil and mousse microcosm experiments, and detected no removal of nC-17 or nC-18 relative to pristane and phytane.

During the meeting of the <u>Researcher/Pierce</u> Cruise Synthesis Steering Committee in February 1981, it was postulated that the presence of living or dead microbial biomass on the surface of mousse flakes may have enhanced agglomeration stability, and may explain the observed increase in non-chromatographable residual materials noted by Boehm and Fiest (1980b) in many of the mousse samples. Atlas et al. (1980) indicated that microbes associated with the mousse could slough off during incubations. However, they also found that microbial degradation of mousse from the IXTOC spill was extremely slow, and that only 7% of the mousse was actually mineralized during 180 day incubation periods and no changes in nC-17/pristane and nC-18/phytane ratios were observed for periods up to 120 days. Nutrient limitation was suggested as the primary factor inhibiting microbial utilization. Only in specialized micro-environments (for example, where mousse was associated with plant material) did microorganisms appear to degrade mousse particles at higher rates; this was due presumably to the availability of nutrients from the decaying plant material (Boehm and Fiest, 1980b). Similar observations were reported previously by Blumer et al. (1973) for stranded tar balls associated with decaying seaweed in the upper intertidal zone.

Fate of Stranded IXTOC Mousse Along the Southeast Texas Coastline

Gundlach et al. (1981) described the persistence of several large patches of mousse observed on the south Texas coast. During August and September 1979, large patches of mousse washed ashore along much of the south Texas coastline. In addition, 20 miles of North Padre Island and four miles of Brazos Island sustained heavy coverage by beached mousse. The oil that reached the shorelines during this period was estimated at 3,900 tons. Following storm activities, the shoreline became noticeably cleaner because sand was deposited over the surface oil. Desiccation of mousse to tar caused a reported reduction to at least half the original volume. By September 3-6, approximately 31% of the beached oil was on the beach surface, 53% was buried, and 16% remained within the swash zone. Thus, even though the beach surface appeared cleaner, the actual mousse content remained approximately the same.

On September 13, very high tides, strong onshore winds, and 1 to 2 m waves were generated by tropical storms. Within two days, over 90% of the remaining oil was removed from the shoreline. Gundlach et al. observed a considerable amount of sheen in the surface swash zone as waves reworked the sediments; however, the small amounts of oil that remained were found primarily high on the beach along the base of a foredune ridge. The sample of beached mousse from Laguna Madre, described above, was collected in this high beach zone region, and the chemical composition was remarkably similar to mousse flakes obtained only 13 to 16 miles from the IXTOC wellhead. After the mid-September tropical storms, at least 36 "tar mats" (deposits of mousse and sediment) were discovered. It was believed that these tar mats were formed from the large oil/mousse masses which were then compressed and mixed with sediment during the tropical storm.

Seven months later, during mid-April 1980, only 19 of the tar mats were still visible, and two different types of mats could be differentiated by color, sediment, and vegetation content (Sadd, 1980). The largest mat was 65 m long, 7.5 m wide, 20 cm thick, and contained 7% oil, 14.7% water (showing significant desiccation), and 77.6% sediment with a specific gravity of 1.38. The amount of oil incorporated in the 19 measured tar mats was estimated at 180 tons, or less than 5% of the maximum quantity of oil found on the Texas shoreline during September 1979. A final beach survey was undertaken in August 1980, a week after the passage of Hurricane Allen. The 19 tar mats remained and, because of the erosion of sand during the storm, several new, though generally very small, tar mats were also uncovered. The ability of these mats to endure Hurricane Allen suggests that they might persist for several more years.

Deposition of oil was found to vary with beach type. Oil buried along shell beaches penetrated to depths up to 40 cm, whereas on fine-grained, sandy beaches it reached a maximum depth of only 7 cm. Smaller patches of mousse were observed to be broken up into tar balls by storm activities, and these were commonly found along the bottom within the shoreline break and swash zone inshore of the first sediment sandbar. The tar balls rapidly decreased in size with distance from shore, from 1 to 5 cm to less than 1 cm, and sediment laden or "armored" tar balls appeared to be the most common at the shorebreak and on the first sandbar. Seaward of the first sandbar, tarballs became very sparse, but evidence of large quantities of nearshore bottom oil was found. Within the first trough, sediment laden mousse patches up to three meters by four meters in size and 10 cm thick were sparsely scattered.

Bedinger and Nulton (1982) also characterized tar samples from south Texas beaches in May, 1980, approximately eight months after the IXTOC oil first washed ashore. The authors described two tar "types": a "soft" tar and a "hard" tar. The soft tar was dull brown to black, 0.5 to 5 cm in diameter, and contained considerable quantities of sand. The hard tar consisted of small flecks up to 10 cm, with a shiny black inner surface, but no sand or inclusions. Furthermore, the hard tar appeared to be more weathered than the soft tar, and Bedinger and Nutton suggested that this material was formed at sea. In contrast, the soft tar appeared to be derived from mousse, mixed sand, flotsom, and small organisms.

In conclusion, more than 476,000 tons of oil were released by IXTOC-I into the Gulf of Mexico, but less than 0.8% washed ashore in Texas. After the storms one year later, less than 0.04% of the oil remained on the beach as tar mats.

BURMAH AGATE

The tanker Burmah Agate and the freighter Mimosa collided near a sea buoy in the Galveston, Texas ship channel on November 1, 1979. Both vessels were engulfed in flaming crude, and large quantities of a light (39.3° API gravity) Nigerian crude oil were spilled from the Burmah Agate. A heavier 21.9° API gravity Nigerian crude held in the Burmah Agate was not believed to be spilled. The spill was fairly continuous for the first few days, and then intermittent flow continued until January 8, 1980 when the fire was extinguished. An estimated 263,000 barrels of the original 400,000 barrel cargo were spilled or burned, and an estimated 630 barrels of residue remained in the burned out tanks after the fire was extinguished. The initial slick, with a width up to 500 ft. and a thickness of 0.1 inches, moved 12 miles in approximately 12 hours (Kelley et al., 1981). Three days after the spill, the slick formed a Z configuration under the influence of estuarine currents. The oil pooled against the ship's hull in patches extending the length of the vessel while still being quite fluid. On the fourth day following the spill, the wind direction changed and the oil moved closer to the shore and formed wind rows parallel to the wind before impacting the shoreline at Galveston. The fluidity of the oil and the observed changes in the shape of the slick caused by the wind and tidal current patterns were evidence of the absence of mousse formation.

Despite the large size of this oil spill, mousse formation was not observed or believed to be a potential problem. Results from previous laboratory studies suggest that a light Nigerian crude would probably not form a viscous, stable mousse, although Berridge et al. (1968 a,b) reported that

Nigerian light could form a borderline (marginal) emulsion (see Table 1). No direct references to the presence or absence of mousse were cited in several reports on this spill, and observations of the slick behavior suggest that mousse formation was not a problem in this incident.

ALVENUS

On July 30, 1984, the tanker Alvenus developed a crack in its No. 2 tank and eventually lost approximately 45,000 barrels (1.9 million gallons) of two Venezuelan crude oils (Merey crude and Pilon crude) about 11 miles southeast of Cameron, Louisiana. The resulting oil slick divided into three patches which were eventually pushed by winds onto the Texas shoreline. Once the oil impacted the beach five days after the spill, there was little evidence of an oil slick in waters off Galveston, although scattered wind rows containing individual tar balls with sizes of approximately 7-8 inches by 3-4 inches by 1-2 inches thick, were encountered. Wind rows were up to four meters across and extended for distances of hundreds of meters. These tar balls were extremely viscous, with a density of 0.94 g/ml and a consistency similar to roofing tar, but did not appear to be emulsified. Furthermore, they were not sticky and did not adhere to the vessel hull. Grabs for bottom sediments and shrimp trawls from several areas along the track of the surface slick did not contain oil, indicating that spilled oil from the Alvenus did not sink in the offshore areas.

The majority of the oil from the Alvenus rapidly stranded along an 18 mile stretch of sand beach on Bolivar Peninsula and on Galveston Island, with the heaviest oiling occurring at beaches south of Galveston. Once the oil reached shore, it appeared to accumulate sediment and, in some places, formed water-in-oil emulsions. In some areas the oil washed into the surf zone and settled to the bottom, forming large continuous tar mats up to three inches in thickness that extended into water depths of two feet. These tar mats were sufficiently heavy that the material did not refloat with wave or tidal current turbulence, although fingers or protrusions of oil would occasionally fold over on itself with passing or breaking waves. Some of the oil washed up onto higher beach elevations and eventually dried into an asphalt-like material as the incorporated seawater evaporated. In other areas where the pools formed above the intertidal zone, the oil remained fairly fluid, considering the high viscosity and residuum content of the starting crude, which was probably due to heating from the sun.

An oil sample collected from the surf zone by Payne and McNabb (unpublished data) had a density of 0.93 g/ml, contained 16% water (by weight), and had a viscosity of 52,000 cP. The aliphatic fraction of the beached oil was devoid of all n-alkanes with boiling points less than nC-12, and the aromatic fraction contained no compounds with molecular weights less than 2-methynaphathalene. The starting Merey crude and Pilon crude had measured viscosities of 230 cP and 2,200 cP, respectively. The Merey crude contained considerably more resolved lower molecular weight compounds, especially xylenes and other alkyl-substituted benzenes, than the starting Pilon crude. Results from an oil weathering model, using the physical/chemical characteristics of the starting Merey and Pilon crudes for model initiation, predicted that a maximum of 13% by weight of the fresh oil would evaporate and only negligible amounts of the oils would disperse into the water column. Under the specified environmental conditions of 84.2°F air temperature and 20 knot wind speed, model-predicted percent water content, viscosity, density, and evaporative losses were in excellent agreement with measured properties for the beached oil (Payne and McNabb, unpublished data).

OPEN OCEAN FIELD TESTS OF SPILLED PETROLEUM

Several controlled open ocean spills have been conducted to determine the chemical and physical behavior of crude oil in the early minutes and hours after the oil is released (JBF/API, 1976; McAuliffe et al., 1981). Approximately 440 gallons of a light and heavy crude oil were spilled separately in four studies in both calm and rough seas in October and November 1975, approximately 75 miles east of Boston. A Murban crude (specific gravity, 0.83; 39.0° API) and La Rosa crude (specific gravity, 0.91; 23.9° API) were used in the studies. While the Murban crude initially appeared to spread faster than the La Rosa crude, the latter may have achieved a larger total slick area because of greater surface slick persistence. Both oils were observed to form lenses or globs of thicker portions of the oil at the leading (down wind) edge of the slick. Evidently these thicker masses maintained a higher wind resistance and had a different frictional resistance at the oil-water interface compared to the thinner slick and sheen which trailed behind. Hollinger and Mennella (1973) have reported that as much as 90% of the total oil may exist in such thick patches in open ocean spills. While the lenses appeared immediately after the spill, "mousse" formation appeared to be delayed by 60 to 90 minutes, and then only occurred under specific conditions. In these studies, mousse was reported to form with the less dense Murban crude, but not with the La Rosa. Surprisingly, mousse formation was only mentioned specifically for the Murban crude, whereas the Venezuelan crude was expected to form a more stable emulsion because of its higher specific gravity. Additionally, results from several laboratory studies have shown that

Venezuelan crudes (in general) are capable of forming stable mousse. As in the IXTOC spill, it appeared that mousse formation only occurred after a sufficient period of time for evaporation and dissolution of lower molecular weight components.

One of the most important results from these studies was the recommendation that time-variable factors for the behavior of thicker lenses and mousse be incorporated into mathematical and descriptive models for more accurate predictions of oil weathering and oil trajectories. Specifically, it was noted that the non-homogeneity of the oil slick must be considered when predicting oil spill transport and spreading behavior. Such approaches have been undertaken by Berridge et al. (1968b) and, more recently, by Grose (1979). Further, it was noted that the chemical differences in terms of composition and weathering rates between the lenses and the thinner oil films should be established. Approaches for studying these differences in chemical properties were undertaken by Overton (1980b) and Boehm and Fiest (1980b) for the IXTOC spill.

In 1978 and 1979 four additional tests were conducted off New Jersey and off southern California (McAuliffe et al., 1981). The 1978 east coast studies consisted of four 10-barrel spills using Murban and La Rosa crudes. In these instances, however, the slicks were sprayed by helicopter with a self-mixing dispersant after two hours. Following dispersant treatment, the Murban crude almost completely dispersed, and the La Rosa crude partially (approximately half) dispersed. Concentrations of oil up to 18 mg/liter at 1 meter in the water column greatly exceeded those concentrations previously observed for the 1975 spills that were subject to natural dispersion.

In controlled oil spills conducted in September 1979, 10 to 20 barrels of Alaskan Prudhoe Bay crude oil were discharged off Long Beach, California. Different approaches to dispersant application were incorporated into the experiments, and untreated slicks were used as controls. Three slicks were sprayed with a self-mixing dispersant from a DC-4 aircraft, three slicks were sprayed with the same dispersant from a boat, and one slick was sprayed with a second dispersant from a boat. Chemical analyses showed that 45 to 80% of the oil was dispersed by aerial treatment. The differential lens thickness (thicker oil patch) phenomenon noted in the 1975 spills was specifically addressed with regard to its behavior following treatment with dispersants. When lenses of thicker oil were treated from the surface vessel, 60% of the oil could be successfully dispersed. In contrast, only 5 to 10% of the oil dispersed when the entire slick was uniformly treated. In the southern California spills it was also noted that the chemically dispersed oil had a weaker tendency to adhere to solid surfaces such as bird feathers, rocks, sand, and sediment present in the water column. Effects of delayed application of dispersant were also tested to evaluate efficiency of dispersants on a more dense and viscous weathered oil. In this test the partially weathered oil was not as effectively dispersed (45% vs 60%). Similar results were obtained from laboratory tests of dispersants with both fresh and artificially weathered Prudhoe Bay crude oil. In the laboratory tests the fresh Prudhoe Bay crude oil had an API gravity of 26.6° API (0.90 g/ml) and a viscosity of 183 centistokes at 15.6°C; the artificially weathered crude (23% volume nC-11 and less removed by distillation) had an API gravity of 22° (0.92 g/ml) and a viscosity of 1,210 centistokes.

In general, dispersants sprayed from the aircraft more effectively dispersed Prudhoe Bay crude oil than when applied from a spray boat. The dispersants worked best when applied neat, as dilution with water destroyed the "self-mix" property that causes dispersion of fine oil-in-water droplets. The amount of dispersant applied also affected removal of oil from the water surface. In 1978, aerial application of 4.9% dispersant resulted in 78% dispersion of the oil, whereas a 3.6% dispersant application yielded only 60% dispersion. In all cases, the best results were obtained when the thicker portions or lenses of the slicks were treated directly. It was not stated in any of the articles whether this thicker material was a true "mousse", but the data clearly demonstrated that dispersants must be applied to the heaviest part of the oil slick for the best results. Dispersants applied to thin oil patches or surface sheens greatly overtreat them while not affecting the thicker oil. When the entire slick surface was sprayed uniformly, only 5 to 8% of the slick dispersed as opposed to 62% when the spray was concentrated on the thicker portions or lenses.

Smith (1977) reported the results obtained from several planned oil spills using #2, #4, and #6 fuel oils, a light crude oil, and a heavy crude oil. After the spills, all of the oils moved parallel to the wind direction and formed a tear drop shaped slick similar to that described above for the Murban, La Rosa, and Prudhoe Bay crude oil spills. The tear drop profile contained the thicker portion of the oil concentrated near the down-wind, leading edge of the slick, with a long tail of increasingly thinner oil extending towards the point of release. This was observed for all oil types and all wind conditions, but was more noticeable in the low to mid-range turbulence conditions where the thin tail was not dispersed as rapidly into the rough seas. This behavior again documents the importance of incorporating time-variable and area-variable slick thickness parameters into algorithms for modeling oil weathering behavior. In general, the thicker oil slicks exhibited a greater leeway and moved faster than the thin slicks. The apparent order of leeway, arranged in decreasing magnitude (for a given wind speed), was heavy crude oil > light crude ≈ #2 fuel oil > #6 fuel oil ≈ #4 fuel oil. Interestingly, there was no apparent systematic correlation between either viscosity or density with oil slick leeway. The oils, arranged in terms of decreasing viscosity and density: heavy crude oil ≥ #6 fuel oil >#4 fuel oil > #2 fuel oil ≈ light crude oil, did not correspond to the observed order of leeway rates.

The weathering rates of the thin vs. thick slicks also appeared to differ, as evaporation of volatile constituents occurred more rapidly from the thinner slicks. Thus, the thin films were enriched (with time) in the higher molecular weight (less volatile) constitutents, including a number of non-volatile constitutents with some surface activity. One explanation for observed behavior of the thin films that trail the leading edge of the slick was that they had surface active materials which tended to make them associate with the water surface and spread to form a mono-molecular film. The results of dye marker studies suggested that the thinner tailing portions of the film had little or no net velocity relative to the water surface because of the presence of surface active materials. The thicker portions of the oil present a larger drag profile to the wind, and exhibit a positive velocity with respect to the surface water. The lens also folds over on itself because of the drag profile of the water. As in the IXTOC I case, and in laboratory and field (wave tank) studies using Prudhoe Bay crude oil, this folding behavior leads to enhanced water uptake and apparent mousse formation in the absence of other, more vigorous, mixing conditions.

CONCLUSIONS

From discussions of the various case histories of oil spill events and controlled studies it is apparent that mousse formation is dependent upon the oil type. Furthermore, field observations of mousse behavior tend to parallel laboratory results, with the exception of the behavior of the Bunker C oil from the US/NS Potomac off Greenland. However, this spill occurred in extremely calm seas, and the absence of emulsion formation exemplified the important role played by physical oceanographic conditions. Results have also shown that the optimum particle size of water droplets incorporated into the more stable water-in-oil emulsions in laboratory mousse formation and field observations of stable mousse appear to be identical.

Finally, it is important to note the drastic differences in environmental impact in open ocean versus coastal spills. In most of the open ocean spill events, where shorelines were not directly oiled (such as the Ekofisk Bravo), the overall impact of even major oil spills has been minimal or short-lived. When direct oiling of coastlines by oil or mousse occurs, deleterious effects may last for 10 to 20 years or more (depending on the energy regime of the coastline). In the case of the IXTOC blowout, generally considered to be an open ocean spill, the magnitude and release time involved were sufficiently large to classify the spill as one exhibiting impact-effects for both open ocean and coastal situations. Thus, while the loss of hydrocarbons at the wellhead from dispersion, dissolution, dilution, evaporation, and eventual sedimentation tended to minimize long-term effects in that area, localized and heavy mousse contamination of impacted estuaries, lagoons, and higher intertidal zones of beaches caused damage which may persist for many years.

CHAPTER 4

TAR BALL FORMATION AND DISTRIBUTION

GLOBAL DISTRIBUTION OF PELAGIC TAR

Numerous reports have appeared during the last decade on the occurrence of tar balls in surface water samples (Horn et al., 1970; Butler et al., 1973; Morris and Butler, 1973; Wong et al., 1973; Sleeter et al., 1974; Mommessin and Raia, 1975; Butler, 1975a, b; Sleeter et al., 1976; Smith, 1976; Wade et al., 1976; Pequegnat, 1979; Shaw and Maples, 1979; Geyer, 1981; van Dolah et al., 1980). Data from these and other studies were compiled by Clark and MacLeod (1977); Table 7, which is adapted from Clark and MacLeod, presents an updated summary of documented tar ball measurements in the worlds oceans as of 1985. From these data it is evident, as noted by Feldman (1973), that the Mediterranean Sea is considerably more contaminated with tar balls than the Sargasso Sea or the North Atlantic. Tar ball concentrations in the North Atlantic and North Pacific were estimated in 1973 to be approximately equal, although Wong et al. (1973, 1974) showed that there was a significant difference in the average levels of tar balls in the western and eastern North Pacific. Tar balls ranging in size from 0.05 to 3 cm were obtained in 30 of 37 tows in the North Pacific. The consistency of these tar balls ranged from hard to soft and tacky, although the majority of tar balls were characterized as hard. A longitudinal center-line of 172.5° W appeared to separate the distinct tar ball distributions in the western Pacific from those in the eastern Pacific; average tar ball concentrations in the western Pacific were 3.8 mg/m^2, whereas the eastern Pacific contained on an average 0.4 mg/m^2. As noted above, this range is similar to that found in the North Atlantic. It has been noted that higher levels of tar balls on the sea surface occur either along ocean routes that support high densities of oil tanker traffic or in areas downstream from these routes (Clark and MacLeod, 1977; NAS, 1975; Butler et al., 1973; Morris et al., 1975).

The distribution of tar balls is highly variable, and concentrations may vary by a factor of 10 or more at a single station during the course of a single day and by as much as a factor of 500 in the course of a year. Cordes et al. (1980) suggested that greater consistency among surface samples for estimating tar ball concentrations could be achieved by towing sampling equipment perpendicular to the wind direction. Despite the variability associated with previous tar ball measurements, estimates of standing crops of tar balls have been made for several of the world's oceans: Northwest Atlantic marginal sea, 2,000 metric tons; Gulf Stream, 18,000

TABLE 7. Summary of Tar Ball Distributions and Concentrations in the Worlds Oceans (adapted from Clark and MacLeod, 1977)

Description of Tar Samples: Ocean, Geographic Area, and Season		Amount of Tar Residue (mg/m^2): Maximum	Mean	Reference
Atlantic Ocean				
Scotia shelf		2.4	0.9	Morris, 1971
Lat. 38^0 to 42^0 N, Long. 50^0 W		9.7	2.2	Morris, 1971
Virginia to Cape Cod, Coastal	Winter	4.4	1.04	Sherman et al, 1973; Sherman et al, 19
	Summer		0.18	Sherman et al, 1973; Sherman et al, 19
	-	0.2	0.04	Attaway et al, 1973
Offshore	Winter	11	0.05	Sherman et al, 1973; Sherman et al, 19
	Summer		0.77	Sherman et al, 1973; Sherman et al, 19
North Carolina to Florida	Winter	-	1.22	Sherman et al, 1973; Sherman et al, 19
	Summer	-	0.23	Sherman et al, 1973; Sherman et al, 19
	-	20	5.5	Attaway et al, 1973
North Antilles & Bahamas	Winter	87	4.8	Sherman et al, 1973; Sherman et al, 19
	Summer		3.9	Sherman et al, 1973; Sherman et al, 19
Lesser Antilles		8.37	1.12	Sleeter et al, 1976
Ocean Station BRAVO, Labrador Current		0.003	0.00	McGowan et al, 1974
Ocean Station CHARLIE, North Atlantic		1.83	0.12	McGowan et al, 1974
Ocean Station DELTA, North Atlantic		10.73	1.15	McGowan et al, 1974
Ocean Station ECHO, Sargasso Sea		21.62	2.64	McGowan et al, 1974
Off Bermuda		-	0.6	Morris, 1971
Northeastern North Atlantic		480	-	Butler et al, 1973; Horn et al, 1970
		14.2	4.8	Sleeter et al, 1974
		1	0.6	Attaway et al, 1973
Barents Sea		3.0	0.15	Smith, 1976
Norwegian Shelf		0.4	0.04	Smith, 1976
Northern North Sea		0.2	0.02	Smith, 1976
Skagerrak		12.1	0.32	Smith, 1976
Central eastern North Atlantic		22.6	9.8	Sleeter, et al, 1974
Gulf Stream		0.8	0.3	Attaway et al, 1973
		6.7	3.8	Sleeter et al, 1974
Sargasso Sea		40	9.4	Morris & Butler, 1973
		1.4	0.2	Sherman et al, 1973; Sherman et al, 19
		6	3	Attaway et al, 1973
		90.6	25	Sleeter et al, 1974
South Atlantic Bight				
Lat. 27^0 to 34^0 N, Long. 77^0 to 80^0 W				
	Winter, 1973		0.31	van Dolah et al., 1980
	Spring, 1973		0.46	van Dolah et al., 1980
	Autumn, 1973		0.93	van Dolah et al., 1980
	Spring, 1974		0.83	van Dolah et al., 1980
	Summer, 1974		0.76	van Dolah et al., 1980
	Winter, 1975		1.95	van Dolah et al., 1980
	Off Georgia/Florida, 1979	5.6	0.82	Cordes et al., 1980
Central Atlantic				
Canary Current		7.69	2.02	Sleeter et al, 1976
North Equatorial Current		0.27	0.16	Sleeter et al, 1976
Equatorial Counter Current		0.04	0.02	Sleeter et al, 1976
South Equatorial Current		0.57	0.11	Sleeter et al, 1976
Equatorial Current Region		63.6	12.7	Polikarpov et al, 1971

TABLE 7 (cont.)

Mean, Geographic Area, and Season		Maximum	Mean	Reference
Carribbean Sea		1.5	0.4	Jeffrey, 1973
		0.9	0.2	Sherman et al, 1973; Sherman et al, 1974
		4.5	0.74	Jeffrey et al, 1974
		13.4	1.62	Sleeter et al, 1976
Puerto Rico, SW Coast		10.5	1.0	Corredor et al., 1983
Gulf of Mexico	Offshore	45.3	1.6	Van Vleet et al, 1983, 1981
West Florida Shelf		-	0.05	Van Vleet et al, 1983, 1981
		10.0	1.2	Jeffrey et al, 1974
		3.5	1.1	Jeffrey, 1973
		6.0	1.12	Sleeter et al, 1976
Loop Current		45.3	0.89	Van Vleet et al., 1984
South Texas 1976				
	Winter (Jan.-Feb.)		.99	Pequegnat, 1979
	March		2.25	Pequegnat, 1979
	April		11.21	Pequegnat, 1979
	Spring (May-June)		0.78	Pequegnat, 1979
	July		0.12	Pequegnat, 1979
	August		0.27	Pequegnat, 1979
	Fall (Sept.-Oct.)		0.62	Pequegnat, 1979
	November		0.18	Pequegnat, 1979
	December		5.92	Pequegnat, 1979
South Texas 1977				
	Winter (Jan.-Feb.)		1.26	Pequegnat, 1979
	March		11.11	Pequegnat, 1979
	April		0.41	Pequegnat, 1979
	Spring (May-June)		0.93	Pequegnat, 1979
	July		0.46	Pequegnat, 1979
	August		0.08	Pequegnat, 1979
	Fall (Oct.-Nov.)		1.38	Pequegnat, 1979
	November		1.76	Pequegnat, 1979
	December		0.56	Pequegnat, 1979
Overall average from 2 year study			1.66	Pequegnat, 1979
Mediterranean Sea		540	20	Horn et al, 1970; Morris, 1971
Ionian Sea	1969	540	130	Horn et al, 1970; Morris et al, 1975
	1975	110	16.0	Morris et al, 1975
Alboran Sea	1969	10	6.5	Horn et al, 1970; Morris et al, 1975
	1975	45	11.0	Morris et al, 1975
Tyrrhenian Sea	1969	20	1.5	Horn et al, 1970; Morris et al, 1975
	1975	15	3.2	Morris et al, 1975
Balearic Sea	1969	10	2.4	Horn et al, 1970; Morris et al, 1975
	1975	1	0.5	Morris et al, 1975
Central	(tarballs)	6.1	-	Morris and Culkin, 1974
Eastern	(tarballs)	10.0	4.1	Morris and Culkin, 1974
Central	(emulsions)	0.30	-	Morris and Culkin, 1974
Eastern	(emulsions)	0.36	0.14	Morris and Culkin, 1974
Arabian Sea		31.1	1.1	Benzhitskiy, 1981
Indian Ocean/Cape of Good Hope		232	<0.1	Shannon et al., 1983
Southwest Pacific Ocean				
Lat. 35° N, Long. 140° E to 175° W		14	3.8	Wong et al, 1974
Lat. 25° to 40° N, Long. 140° to 160° W		16.3	2.1	Wong et al, 1976
Outside the Kuroshio Current		-	0.4	Wong et al, 1976
Southeast Pacific Ocean				
Lat. 35° N, Long. 175° to 130° W		3	0.4	Wong et al, 1974
Lat. 25° to 40° N, Long. 140° to 160° W		-	0.03	Wong et al, 1976
South Pacific Ocean		-	0.0003	Wong et al, 1976

metric tons; Sargasso Sea, 66,000 metric tons; total Northwest Atlantic, 86,000 metric tons; Mediterranean, 50,000 metric tons; total Northwest Atlantic and Mediterranean, 136,000 metric tons (Morris & Butler, 1973; Feldman 1973). The extreme variability in the data and lack of synoptic and sufficient measurements from broad expanses of the world's oceans, make such standing crop estimates tenuous at best. At this time it is impossible to determine whether tar ball concentrations are increasing or decreasing. Further, Wade et al. (1976) stated that many small particles (ranging from 0.3 μm to 1.0 mm in diameter) present in surface waters were not sampled routinely during previous studies. Therefore, previous estimates of pelagic tar levels may be too low because they did not include the smaller size range particles. Morris et al. (1976) measured tar-like particles with diameters of 10 to 500 μm in the Sargasso Sea and estimated that their total mass in the water column to a depth of 100 m was about four times greater than the standing crop of larger pelagic lumps at the ocean surface. These smaller particles would be more apt to disperse into the water column and would not contribute to pelagic tar; therefore, including these particles in surface tar ball estimates is probably not appropriate.

CHEMICAL COMPOSITION OF TAR BALLS

Butler (1973) provided a detailed summary of information on the distribution and chemical composition of tar lumps found on the surface of the ocean prior to of 1973. Balkas et al. (1982) used a series of analytical techniques, including IR, GC, GC/MS, and H-nmr, to characterize chemical differences between floating and sunken tar balls in the Mediterranean. Gas chromatographic profiles of a number of samples showed that, in general, aliphatic and aromatic materials with molecular weights less than nC-14 to nC-17 were depleted, and most of the samples showed an evenly repeating series of alkanes out to nC-45. Bimodal distributions of lower and intermediate molecular weight alkanes were observed in several instances, and unresolved complex mixtures characterized many of the samples.

Koons (1973) analyzed 34 tar balls collected from Texas beaches and from the Gulf of Mexico. The physical appearance and chemical composition of these samples varied considerably. Tar balls ranged in size from a few millimeters to several centimeters, and some samples were quite soft and appeared to flow on the beach due to heating from the sun, whereas others were quite hard, almost brittle, and could be broken with a clean fracture. The saturated hydrocarbon composition of the tar balls ranged from 1.6 to 56.1%, and the asphaltene content ranged from 8.8 to 54.7%. Average values of the 34 samples were as follows: saturated hydrocarbons, 31%; aromatic hydrocarbons, 24%; LC-elutable NSO's, 14%; non-LC-elutable NSO's, 6%; and asphaltenes, 25%. Evaporation and dissolution had stripped away hydrocarbons up through nC-15 to nC-17 in all of the samples, and in the fresher appearing samples, ratios of nC-17 to pristane and nC-18 to phytane were not substantially different from those of the seep oils from the area. The harder, brittle tar ball samples had undergone appreciable chemical and biological oxidation, in addition to physical evaporation/dissolution weathering. These latter samples generally contained lower levels of saturated hydrocarbons (1.6 to 10%), but greater amounts of NSO's and asphaltenes. Further, the ratios of saturate to aromatic hydrocarbons were significantly lower (by a factor of 10) than those for the seep oil and the nC-17/pristane ratios approached zero. Both of these observations suggest that appreciable biological degradation of the tar ball samples had occurred.

Mommessin and Raia (1975) summarized chemical characterization data on a total of 110 tar samples collected from the northwestern Atlantic, the Sargasso Sea, along shipping routes in the vicinity of Coast Guard Stations Echo and Delta in the Northwest Atlantic, along the Florida coast, in the vicinity of New York harbor, along the Texas coast near Galveston, and from three samples collected near Coast Guard Station November in the Pacific Ocean. The tar ball samples ranged from pea size (1 cm) to 3 to 7 cm in diameter. In general, the tars were black or brown/black and had irregular shapes, and the texture varied from very hard and dry to soft and sticky, although three of the samples examined were essentially liquid. Surface portions of a number of samples were shiny with substantial oil stain. As noted in the case history of the IXTOC blowout, reseachers observed numerous large tar and mousse balls which contained accumulation of fresh oil on the surface, demonstrated by a darker black oil stain.

Several of the tar balls examined by Mommessin and Raia were coated with tree leaves, Sargassum, unidentified plants, barnacles, bryozoans, and other pelagic organisms. In other samples, the presence of insects and wood fragments, in addition to terrestrial plants and leaves with substantial quantities of quartz and carbonate sand or clay particulates, suggested that the tar balls may have been beached at one time. In the IXTOC spill, numerous large tar balls and mousse agglomerations were observed with sugar cane stock or other intertidal debris present. These substrates were believed to act as sites for nucleation and further mousse accumulation. Mommessin and Raia also noted that pieces of plastic, bits of buttons, tubing, metal fragments, synthetic fibers, fly ash, and rope were frequently abundant, illustrating that industrial and commercial byproducts are a source for numerous tar samples.

The inside portions of the tars in the Mommessin and Raia study did not contain materials boiling below nC-12, whereas residues boiling above nC-34 were very high; similar observations were reported by Blumer et al. (1973) and Butler (1973). Distinct variations were observed for the semi-volatile materials lower than nC-34, and of the 110 samples analyzed, the median value for residual materials with molecular weights greater than nC-34 was 64%; 90% of the samples analyzed had residues of the nC-34+ material ranging between 38 to 80%. In a number of the samples, large peaks were observed at nC-20 and nC-23, and unlike refined or crude petroleum products which show a smooth distribution curve throughout the distillation range, the presence of these compounds suggest biogenic material or manufactured chemicals. Most of the samples containing these compounds came from New York Harbor and were believed to be influenced by urban and/or industrial activities. Only two of the samples had appreciable levels (5 and 12%) of lower molecular weight compounds between nC-10 and nC-30; however, the n-C34+ residue contents were also highest in these samples (70 and 80%, respectively). The presence of the low boiling components was taken to suggest that the samples had a relatively short history in the marine environment.

The 110 samples were organized into five groups depending on the molecular weight distributions of the alkanes observed. Tar, asphaltene, and resin characterizations were performed on the samples. Sulfur contents in the tar ranged from 0.1 to 2.7%. The asphaltene fractions generally had higher levels of sulfur, with percentages in most samples in the range of 1 to 3.5%. The resin fractions contained sulfur compounds in approximately the same range as the asphaltenes. Infrared spectra of the tar and asphaltene fractions were similar to those for petroleum derived waxes and oxidized petroleum products, and suggested the presence of oxygenated carbon compounds.

The authors concluded that two distinct sources for the types of tar samples were indicated. One source appeared to be urban and industrial waste products, and the other source included petroleum-based materials modified by various weathering processes. A high sulfur content was noted in tar samples collected from Florida, suggesting that the samples originated from sour crudes known to be transported in that area. The presence of substantial specific components at nC-20 and nC-23, and low sulfur contents with notable carbon double-bond oxygen infrared absorptions, suggested that most of the samples from the New York Bight were derived from waste or manufacturing processes. The higher boiling normal n-alkanes found in many of the open ocean samples were attributed to materials in discharges from tanker washing and ballasting operations. It was interesting to note that in some instances several samples collected from the same vicinity showed similar compositions; whereas, many other samples collected at the same locations showed no correlation between the composition and the sampling location.

Wade et al. (1976) analyzed a number of smaller tar particles (0.3 μm to 1.0 mm diameter) by gas chromatography and infrared spectrometry and found that the pelagic tar samples averaged about 32% water (with a range of 11 to 44%) and 68% dry weight material. Butler et al. (1973) found that tar lumps from the Atlantic Ocean typically contained about 25% water by weight. An average of 53% of the wet tar was soluble in benzene (range: 31 to 89%), and this material accounted for approximately 78% of the average dry weight of the samples. The benzene insoluble fraction of tar possibly included inorganic salts, non-organic debris, and higher molecular weight material. The wet tar samples averaged about 16% total hydrocarbon material, and the remaining weight percentage included non-hydrocarbon organic material or hydrocarbons not detected by their procedures (for example, hydrocarbons less than nC-14 and greater than nC-38). Jeffrey et al. (1973) found that Gulf of Mexico pelagic tar samples contained an average 26% asphaltenes based on the dry weight of sample. The gas chromatographic analyses showed that the predominant hydrocarbon weight percent was from components in the unresolved complex mixture (average: 79%; range: 67-97%) which includes both aromatic and cycloparaffinic compounds. Varying degrees of resolved alkanes were observed in the different samples, with several samples showing no resolved peaks over the UCM and one sample showing an evenly repeating series of alkanes from nC-15 to nC-34. A number of other samples contained alkanes only above nC-25, with evidence of persistent pristane and phytane suggesting microbial degradation of the lower molecular weight normal paraffins. Infrared analyses showed that of the eight samples considered, all but two contained aromatic hydrocarbons. Van Vleet et al. (1983) reported that 40% of the tar ball samples collected in the Gulf of Mexico contained a bimodal distribution of n-alkanes within the ranges of nC-19 to nC-20 and from nC-29 to nC-35; 60% of the samples contained a unimodal distribution of n-alkanes with a maximum from nC-18 to nC-28. The nC-17/pristane values ranged from 0.14 to 4.6, whereas, the nC-18/phytane values ranged from 0.31 to 5.2. These chromatographic characteristics suggested that tar balls originated from multiple sources (e.g., crude oil sludge from tanker washings and refined products from tanker ballast operations), and that the tar balls had been degraded extensively by bacteria, whereas other samples showed very little microbial degradation. Cordes et al. (1980) analyzed tar balls from the South Atlantic Bight which contained approximately 30% polycyclic aromatic hydrocarbons. The primary compound in this fraction was perylene, which the authors suggested was highly resistant to weathering.

The elevated levels of NSO compounds found in a number of tar balls are frequently higher than expected from simple evaporation and dissolution of weathered crude oil. This enrichment could quite easily accompany the formation of NSO compounds from petroleum hydrocarbons by photo- and microbial-oxidation. While these photo-oxidized NSO compounds would have enhanced water solubilities, and might be removed easily from the surface of the tar mass, increased diffusion coefficients as a result of elevated viscosities and specific gravities may prevent the loss of these materials. Also, with semi-fluid tar masses folding over on themselves and aggregation of small flakes during tar ball formation, these materials may be entrained in the interior where ultimate loss would be inhibited by decreased diffusivities in the viscous residues. Further, bacteria have been shown to coat the surface of tar balls (Butler, 1973), and oxidation products, cellular materials, and bacterial cell wall components can contribute to the residual non-chromatographable materials and NSO compounds.

SOURCES OF PELAGIC TAR

It is generally accepted that the waxy crude-petroleum sludge from cargo tank washings of tankers is a major source of much of the pelagic tar in the world's oceans (Butler et al., 1973; Koons, 1973; Berridge et al., 1968a; Blumer et al., 1973). Evidence for this source comes from the high content of ferric oxide (up to 18% on a dry weight basis) in some residues (Attaway et al., 1973; Burns et al., 1982) and the predominance of paraffin-rich wax inclusions which are a common feature in open ocean tar balls (Blumer et al., 1973). The elevated levels of iron are believed to originate from rust particles that are picked up by the oil from steel tanks or apparatus encountered during pumping, storage, and transport of petroleum by ocean going tankers. The paraffin-rich wax inclusions found in many tar balls are believed to result from the waxy precipitates which can form during production and transport of petroleum products.

In general, normal alkanes are relatively soluble and uniformly dispersed through out most oils. With long storage and transport at lowered temperatures, however, these materials can precipitate in a form which adheres to surfaces, causing problems in pipelines and tanks (Blumer et al., 1973). The formation of precipitates during transport in oil tankers is one of the reasons that necessitates tank cleaning. Furthermore, the fact that such waxy precipitates are not present in free flowing oil, in submarine seeps, or in oil from blowouts led Blumer et al. (1973) to propose their presence as an unequivocal characteristic of oil from tankers that either were not equipped for load-on-top procedures or did not follow that loading method. The authors also hypothesized that these waxy aggregates may play an active role in the formation of tar balls, since the higher molecular weight waxes are not readily dispersed and can provide a nucleation site around which additional floating oil may accumulate. Thus, McAuliffe (1977) stated that most of the tar balls stranded on many beaches come from tanker compartment washings, Bunker C discharges, or bilge pumping. He added that most crude oil spills probably contribute only small amounts of oil for formation of tar balls. Field observations from a number of open ocean and nearshore spills have shown that this is not necessarily the case. As described in the previous chapter on Case Histories, tar balls reportedly formed in substantial quantities after the tanker Arrow spill, the Metula spill, Ekofisk Bravo blowout, the Potomac spill, and the IXTOC-I blowout in the Bay of Campeche.

FATE OF PELAGIC TAR AT SEA

Wade et al. (1976) suggest that the ultimate fate of many large tar patches would be fracture and break-up by sea-surface turbulence and agitation. They simulated this process by shaking a 1 cm pelagic tar lump in filtered Narragansett Bay water at room temperature for a period of four weeks. Following the agitation period, the mixture was filtered through a 1 mm mesh screen. Wade et al. found that the larger particles (greater than 1 mm retained the basic gas chromatographic features of the original tar. The smaller particles (less than 1 mm) contained mostly unresolved hydrocrabons, and, in that regard, were similar to extracts of whole-water samples collected in the Sargasso Sea at the time the larger tar balls were collected. Thus, Wade et al. suggested that the major source of hydrocarbons (primarily cyclo-paraffins) which were found in their unfiltered Sargasso seawater samples was from particles of weathered pelagic tar in the size range from 0.3 μm to 1 mm in diameter. Peak and Hodgson (1966, 1967) and Gordon et al. (1973) mixed hydrocarbons with distilled water and found that particles of oil less than 1 micron in diameter increased the whole-water oil concentration to values greatly in excess of those for saturated solutions. This excess above saturation concentrations was due to the presence of colloidal particles. Gordon et al. (1973) also noted that the greatest number of particles formed by mixing hydrocarbons with seawater occurred in the size range of 1 to 30 microns.

Large tar particles have also been shown to support several marine organisms (Mommessin and Raia, 1975; Horn et al., 1970). Benzhitskiy (1981) observed that tar balls collected in surface waters of the Arabian Sea frequently were encrusted with diatoms, blue-green algae, bryozoans, sedentary polychaete worms, and eggs of marine insects. Horn et al. (1970) reported that an isopod (Idothea metallica) 10 to 25 mm in length remained with the tar ball when placed in a bucket onboard a research vessel, and the goose barnacle (Lepas pectinata) was frequently found attached to well-weathered (firm) tarry lumps. At one location, four tar lumps had a total of 150 barnacles from 2 to 8 mm in length. The barnacle growth rate was measured at 1 mm per week, indicating a minimum age for the tar of 2 months (Horn et al., 1970). Increased colonization would be expected to eventually lead to sinking when the buoyancy of the tar ball was exceeded by the weight of attached organisms. As noted in the IXTOC spill case history, and by Mommessin and Raia above, other detrital materials are also associated with tar lumps, and accumulation of detritus can also lead to increasing weight and eventual sinking. High levels of suspended particulate material have been found in tarry residues collected off of Florida (Attaway et al., 1973). Gundlach et al., (1981) noted that sand encrusted tar balls from the IXTOC-I spill were rapidly removed from the mid and lower intertidal zones of several Texas beaches during storm activities. These tar balls were found along the bottom within the shoreline break and swash zone. Seaward of the first sand bars, however, they became very sparse and presumably were dispersed by nearshore currents. Tar balls generated from heavier crudes, or from petroleum products such as Bunker C, can achieve high enough densities with the loss of volatile and soluble components to sink to near-bottom waters. Following a Bunker C oil spill in San Francisco Bay in 1971, one to three cm, near-buoyant oil drops were observed by remote control television moving with bottom currents just off the sea floor in 30 feet of water in Bolinas Bay (Conomos, 1975; McAuliffe, 1977). Thus, most pelagic tars are believed to eventually break up during weathering at sea. These smaller droplets are then dispersed in near-surface waters, and most are believed to remain suspended and be dispersed by water currents (Brown et al., 1973, 1975; Brown and Huffman, 1976).

Microbial degradation of large tar balls probably has only a minimal effect on their fate because oxygen and nutrients do not diffuse into the tar balls at sufficient rates to sustain degradative processes. Bacterial utilization of the smaller dispersed materials would be enhanced by the increased surface area-to-volume ratio, and may play a more important role in their ultimate degradation.

Butler (1975a) used a simple, semi-quantitative model of evaporative weathering to estimate the breakup and residence time of petroleum residues at sea. This model assumed that evaporative losses were proportional to the components equilibrium vapor pressure and the amount remaining in the sample at a given time. It was stated that the equilibrium vapor pressure would not be precisely equal to the vapor pressure of the pure compound because petroleum is a mixture of compounds. However, large variations in the activity coefficients as weathering processes occurred were not anticipated or incorporated into the model. Although Butler noted that as evaporation occurs at the surface and loss of components within the slick becomes limited because of diffusion control, the prospect for diffusion control was not included in the model. Artificial parameters were used to limit compounds larger than nC-16 from diffusing into the surface sheen around the larger tar patches to more closely approximate observed oil spill behavior. A residence time for oil on the order of one year was obtained in earlier studies based on a mass balance approach (combining estimates of the standing stock of tar with estimates of the amount of oil spilled annually and the fraction of this oil which remains as pelagic tar) (Morris, 1971; Butler et al., 1973). This estimate was not consistent with the lifetime predicted by the evaporative weathering model unless the actual diffusion rates were smaller than expected or the lumps collected at sea comprised fragments of the original lumps from which weathering began (Butler et al., 1976). Both of these hypotheses were applicable, but they complicated the quantitative dating of tar lumps. Estimating the lifetime of pelagic tar at sea is complicated further by the observed tendency of larger tar balls to break up into 1 μm to 1 mm fragments which can be dispersed into the water column to depths greater than 100 m.

As noted in the Arrow spill, ingestion of oil particles by copepods can enhance their sedimentation rates. For example, Sleeter and Butler (1982) estimated that grazing zooplankton in the Sargasso Sea could ingest 16 to 46 μg of hexane-extractable hydrocarbons per square meter within a four hour period. Subsequent excretion and sinking of zooplankton fecal pellets may result in a vertical flux of 23 mg hydrocarbons$\cdot m^{-2} \cdot yr^{-1}$, which is the same order of magnitude as the estimated input rate of petroleum residues into surface waters of the Sargasso Sea. The high sinking rates of these fecal pellets (e.g., 300 meters per day; Honjo, 1980) and their greater resistence to microbial degradation, suggest a possible mechanism for enhancing the sedimentation and deposition of particulate petroleum residues. However, it is still unclear at this time whether the ultimate fate of dispersed petroleum residues in the open ocean is biodegradation or sedimentation. From the few sediment samples from the top 5 cm of the deep ocean floor in the Sargasso Sea which have been analyzed, hydrocarbon distributions appear to be a mixture of biogenic and petroleum mixtures. The concentrations of these materials (on the order of 50 mg/m^2) are comparable to particulate concentrations in the water column found near the surface (40 mg/m^2) (Farrington and Tripp,

1975; Butler et al., 1976). It is not possible, however, to directly correlate these observations because of complications from horizontal advection and because the depth in the sediment column from which the samples were obtained and the ratio of petroleum to biogenic hydrocarbons were not known (Butler et al., 1976). Evidence from the Ekofisk Bravo and IXTOC I blowouts also suggests that little of the released oil from either of these events ended up in the sediments near the well sites. But, as noted in the section on Case Histories, these results may have been influenced by the sampling techniques (Smith-McIntyre grab samplers) and by resuspension of bottom sediments due to severe storm activities prior to sampling.

Degradation rates of known oil concentrations in benthic sediments were found to be very concentration dependent (Griffiths and Morita, 1981). Payne et al. (1981a) studied the chemical changes in oiled sediment plots in 15 to 30 meters of water in the sub-arctic environment of Kasitsna Bay, Alaska. Results from these studies indicated that the most rapid weathering of the oil occurred at the air/sea interface or in the water column before the oil was incorporated into the sedimentary regime. In an experiment where both fresh and weathered Cook Inlet crude oil were spiked into the sediments at concentrations of one part per thousand (ppt), n-alkanes were depleted, whereas the losses of higher molecular weight aromatic hydrocarbons over a one year period were negligible. At levels of 50 ppt (with both fresh and weathered crude oil), nearly complete inhibition of abiotic and microbial degradation of aliphatic hydrocarbons was observed. In the samples originally spiked with 0.1 ppt oil, there was little or no evidence of either aliphatic or aromatic hydrocarbon contamination after one year. Sediment plots spiked with fresh and weathered crude oil at 50 ppt plus added nutrients (starch and chitin) showed no evidence of enhanced biotic recovery or selective hydrocarbon utilization. This lack of recovery may have been from the toxic aromatic compounds in the oil itself, rather than from inhibition of microbial activity by limited nutrient concentrations.

FATE OF BEACHED OR STRANDED TAR BALLS

The occurrence of tar ball residues on beaches has been reported on an almost global scale (Morris and Culkin, 1974; Morris and Butler, 1973; Butler et al., 1973; Saner and Curtis, 1974; Sleeter et al., 1976; Wong et al., 1976; Dwivedi and Parulekar, 1974; Attaway et al., 1973; Knap et al., 1980; Burns et al., 1982). Table 8, adapted from Clark and MacLeod (1977), summarizes the tar ball residues observed on beaches from a variety of locations.

TABLE 8. Summary of Stranded Tar Ball Distributions and Concentrations on Beach Surfaces (adapted from Clark and MacLeod, 1977)

Description of tar samples Geographic area	Amount of tar residue (g/m^2)	Reference
Southwest Florida coast, 1 km of shoreline	23	Saner and Curtis, 1974
Bermuda	190 (mean)	Butler et al, 1973
Surf Bay 1978-79	338 (mean)	Smith and Knap, 1985
1982-83	139 (mean)	Smith and Knap, 1985
Whalebone Bay 1978-79	1108 (mean)	Smith and Knap, 1985
1982-83	248 (mean)	Smith and Knap, 1985
Barbados, Windward Shore	40-62	Sleeter et al, 1976
Leeward Shore	4.5	Sleeter et al, 1976
Puerto Rico, North Shore	52-112	Sleeter et al, 1976
West Shore	12-20	Sleeter et al, 1976
Central Caribbean Island, Windward Shore	13-230	Sleeter et al, 1976
Leeward Shore	0-2.2	Sleeter et al, 1976
Hondurus, Windward Shore	90-127	Sleeter et al, 1976
Southern California		
Sunset Beach (Long Beach)	0.018-1.35	Ludwig and Carter, 1961
Torrance (Los Angeles)	0.006-0.92	Ludwig and Carter, 1961
Mussel Shoals (Sea Cliff)	0.003-0.60	Ludwig and Carter, 1961
Summerland Beach (Santa Barbara)	0.002-0.38	Ludwig and Carter, 1961
Coal Oil Point	1.22-23.9	Ludwig and Carter, 1961
Gaviota Beach	0.023-1.01	Ludwig and Carter, 1961
Beaufort Sea, Canada 264 km of shoreline (1974)	No tar residues	Wong et al, 1976
Yukon 16 km of shoreline (1974)	380 g grease	Wong et al, 1976
N.W.T. 10 km of shoreline (1974)	425 g grease	Wong et al, 1976
N.W.T. 86 km of shoreline (1974)	157 g grease	Wong et al, 1976
West India Coast	4,480 (max.)	Dwivedi and Parulekar, 1974
Israel Coast	3,625 g m^{-1} shoreline (mean)	Golik, 1982
Oman Coast	224 g m^{-1} shoreline (mean)	Burns et al., 1982

The quantitative distribution of beached or stranded tar at selected locations over time has been proposed as a means to monitor the levels of oceanic tar. Knap et al. (1980) reported the results of quantitative surveys of beach tar on Bermuda during 1971-1972 and 1978-1979. A 15% increase in mean levels for the 1978-1979 data set, while not statistically significant, lead the authors to conclude that during the period of the study the input of petroleum residues to the Sargasso Sea did not decrease, but instead may have increased. Interestingly, this possible increase directly contradicts a reported 27% decrease in operational discharges for tankers during the same period. Recently, however, Smith and Knap (1985) reported that tar levels on selected Bermuda beaches during 1982-1983 were significantly lower than levels observed in 1978-1979 or in 1971-1972. In this case, the authors attributed the reduction in tar strandings to restrictions in operational discharges and a decrease in the number of marine oil spills. The authors also warned that storm waves can uncover old, previously buried, tar balls and redistribute tar within the higher beach elevations. These processes can result in considerable variability in measured tar levels that may obscure any long term trends.

Because of the accessibility of stranded tar balls on beach surfaces, their fate and weathering have been better characterized compared to tar balls dispersed in the water column. Golik (1982) described the following paradigm for the fate and behavior of tar stranded on a beach. Tar balls driven onshore by wind, waves, or currents tend to accumulate sand, shells, and debris that increases the specific gravity of the particles. The heavy, weathered particles eventually are carried further up on the beach by storm waves and high tides, and subsequently accumulate along the swash marks. Tar balls eventually may be buried in the beach up to depths of 50 to 70 cm. However, subsequent storm waves may uncover buried particles and move the particles in an offshore or longshore direction to a site of sediment deposition or to an estuary. Tar particles eventually may break into smaller particles which are buried, mixed with the existing substrate, or further dispersed. As in the case for major oil spills impacting intertidal zones (considered in Chapter 3 on Selected Case Histories), results of the studies on isolated tar ball degradation suggest that weathering rates are also dependent upon the degree of exposure, availability of nutrients, and extent of mixing into the intertidal sediments. Obviously, aside from aesthetic considerations, the environmental impacts of tar ball stranding in intertidal zones are far less serious than those which occur from massive oilings.

Blumer et al. (1973) conducted a detailed study on the chemical changes that occurred in stranded tar in Bermuda and in Falmouth Harbor, near Woods Hole, Massachusetts. In general, evaporation processes predominated, and after 13 to

16 months the oil retained about 10% of the compounds boiling near nC-17 to nC-18 and 50% of those compounds boiling between nC-19 and nC-21. Compounds above nC-23 to nC-24 did not evaporate during the extended time spans. In contrast, oil incorporated into bottom sediments and oil consumed by organisms soon after a spill shows little evaporative loss, even for compounds as volatile as nC-12. Climatic differences between Martha's Vineyard and Bermuda had only minimal effects on the evaporative loss. At the Bermuda station, the oil formed a crust which eventually cracked and exposed additional surface area, which perhaps slightly increased the rate of evaporative loss.

Microbial degradation of n-alkanes and other components in the stranded oil was greatly dependent on the presence of nutrients. Oiled rocks placed in enclosures in the high intertidal zone showed essentially no microbial degradation. In the presence of decaying seaweed in this upper intertidal-zone enclosure, alkanes were depleted within four months. In the absence of decaying plant material in the upper intertidal splash zone of the Bermuda station, microbial degradation occurred at markedly slower rates than at Martha's Vineyard. When microbial degradation did occur, lower molecular weight components were utilized completely before higher molecular weight compounds above nC-25 were degraded. Aromatic and cyclic compounds appearing in the gas chromatographic unresolved complex mixtures suggested that the dissolution process was extremely slow and limited to compounds below nC-20 in the UCM. After 13.5 months of exposure at Bermuda the oil showed little evidence of purely chemical alteration. The initial ratio of saturate to aromatic hydrocarbons was preserved, and the asphaltene and hydroxyl contents did not noticeably increase. A modest increase in ester or acid content was noted; however, after 13.5 and 16 months at Bermuda and Martha's Vineyard, respectively, the spill residues were still far from inert asphalts. They had not been depleted of the more biologically active higher molecular weight aromatics, and the half-life in terms of contamination potential for such beached materials was estimated to be in terms of years.

CHAPTER 5

ALGORITHMS AND COMPUTER PROGRAMS TO SIMULATE THE FORMATION OF WATER IN OIL EMULSIONS

Numerous mathematical formulations have been developed to describe various individual aspects of water-in-oil emulsion behavior including: 1) the work (or mixing energy) required to generate emulsions; 2) the viscosity changes of water-in-oil emulsions as a function of water content; 3) the competitive process of dispersion of oil and mousse into the water column (oil-in-water dispersion); 4) the thickness and spreading of oil and mousse; and 5) the breakup or decomposition of slicks based on evaporation and dissolution weathering and creaming (mousse destabilization by separation of water and oil into distinct phases). Unfortunately, no single oil weathering model exists at this time that encompasses all of these factors. Mackay et al. (1980) have stated that it is impossible to completely model water-in-oil emulsion formation and behavior because of the lack of a sound understanding of the physical chemistry of this system. Perhaps the most pragmatic and successful approach to this problem has been to postulate a realistic mechanism for the emulsification process, and relate the viscosity of the resultant mousse to the water content (Mackay et al., 1980). Algorithms for each of these separate, but interconnected, processes will be considered briefly below.

In citing work by Becher (1955), Twardus (1980) stated that the work or energy required for generation of either water-in-oil or oil-in-water dispersions could be theoretically calculated. Before emulsification occurs, the interfacial area between two immiscible liquids is at a minimum and is essentially determined by the surface tension values of the two liquids. Following emulsification, it is assumed the droplets of the first liquid (A) are dispersed into the second (B). If all droplets are considered, the interfacial area between liquids A and B is increased. Since liquids tend to keep their surfaces to a minimum, an emulsifying agent and work (or energy) are clearly required to allow emulsification to occur. In theory, the amount of work (ω) required to increase the surface area by an amount (S) can be described by the following formula from Becher (1955):

$$\omega = \gamma \Delta S$$

where γ is the interfacial tension between the two liquids. An emulsifying agent, as noted in previous chapters, may be any surface active substance which forms a thin interfacial film between liquids A and B, maintaining the emulsion by minimizing the contact and aggregation of the dispersed liquid. Additional chemical properties of this stabilizing agent are discussed by Twardus (1980) and were considered briefly in the general discussion of emulsification processes in Chapter 1.

Taking advantage of the considerable literature on the rheological properties of suspensions of rigid particles, Mackay et al. (1980) have used the Mooney equations (Mooney, 1951) to develop a relationship between the viscosity of mousse and its water content

$$\mu/\mu_0 = \exp(2.5W/[1-K_1W])$$

where μ is the apparent mousse viscosity, μ_0 is the apparent oil viscosity, W is the fractional water content of the emulsion, and K_1 is a constant. Figure 11, from Mackey et al. (1980), shows a plot of the curves generated with this equation and the fit of experimental data obtained for six crude oils. Although there is some spread in the data, the equation gives a good average fit to the measured viscosities. It was noted by Mackay et al. (1980) that the oils usually exhibited non-Newtonian behavior, causing the viscosity ratio to depend on the shear rate as well as the water content. Similar findings have been reported by Bridie et al. (1980b). As noted by Berridge et al. (1968b), the greatest increase in viscosity for these oils occurs after the water content has reached 50%; before that point many oils have the basic flow properties and appearance as the starting crude.

Mackay et al. (1980) then used these relationships to develop a kinetic expression to describe the processes of water uptake and release. The rate of water incorporation into the oil (I) was postulated to depend on the sea state (S) and the viscosity of the oil as follows:

$$I = K_i S/\mu \quad (m^3/m^2 \cdot s)$$

The rate of coalescence or water removal (R) was estimated to be dependent on the oil composition, its water content (w), viscosity, and the slick thickness (D) as follows:

$$R = K_\tau W/\mu D \quad (m^3/m^2 \cdot s)$$

where K_τ is a coalescing tendency constant.

Thus at any time the net rate of water incorporation would be:

$$d(VW)/dt = A(I-R)$$

where V is the emulsion volume and A is the area. Thus, D is V/A. Substitution yields:

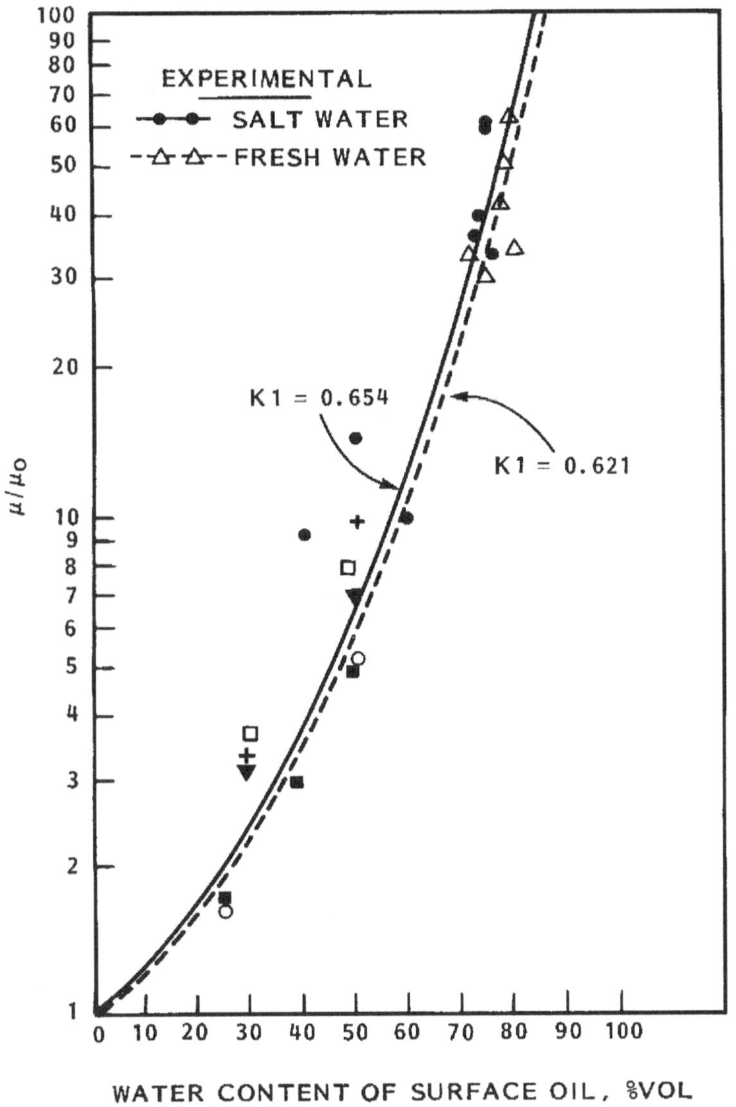

CRUDE NO		SOURCE
●	1	ESCAROS, NIGERIA
+	2	MLB, LIBYA
■	13	BELLVUE, SAN JOAQUIN
▼	8	BREGA, LIBYA
○	14	TEN SECTION, SAN JOAQUIN
□	10	ZVEITNA, LIBYA

Figure 11. MOUSSE/OIL (μ/μ_o) RELATIVE VISCOSITY RATIO OF SIX TEST CRUDE OILS AS A FUNCTION OF WATER CONTENT. (FROM MACKAY et al., 1980.). REPRODUCED WITH PERMISSION OF THE AUTHOR.

$$dW/dt = (K_i S/D - K_\tau W/D^2)/\mu$$

Complete discussions of the derivation of these equations and their relationship to physical properties is beyond the scope of this review; the reader is referred to the original articles by Mackay et al. for details.

The general solutions to the differential equations generated in Mackey's paper were presented; however, explicit solutions could only be provided when the water-in-oil emulsion showed no tendency for coalescence (because of high concentrations of surfactant or coalescence preventing materials) or when the rate of water removal equalled the rate of water incorporation and the viscosity and water content stabilized at a constant value. This limitation applies to oils which do not exhibit a mousse forming tendency. Nevertheless, the equation did have predictive capabilities in generating emulsion formation rates and tendencies for a variety of oils ranging from heavy fuel oils, which exhibit rapid stable emulsion formation, to light distillates, which reject water rapidly and have little or no emulsion formation tendency.

Cormack and Nichols (1977) reported data for water contents for three spills of Ekofisk oil, and fit the data to curves generated with the proposed equation (Figure 12). Very acceptable fits of the observed field data and predicted water uptake for varying wind speeds were obtained. Half times of emulsion formation were generated from these studies, and values of 2.8 hours at 3.1 knots, 16 minutes at 10 knots, and 1.6 minutes at 31 knots were obtained (Mackay et al., 1980).

A simpler finite difference form of the equations used to describe the process of emulsion formation was also presented by Mackay et al. (1980) as shown below:

$$\Delta W = K_A(U+1)^2(1-K_B W)\Delta t$$

where

W is the fractional water content
K_A is a constant
K_B is a constant with a value of approximately 1.33
U is the wind speed
Δt is time.

Until more experimental data are available, it was suggested that this simpler version would be adequate for describing emulsion formation.

In that Mackay et al. (1979, 1980) and others have noted that mousse formation and oil-in-water dispersion are competitive processes for any given oil under certain temperatures

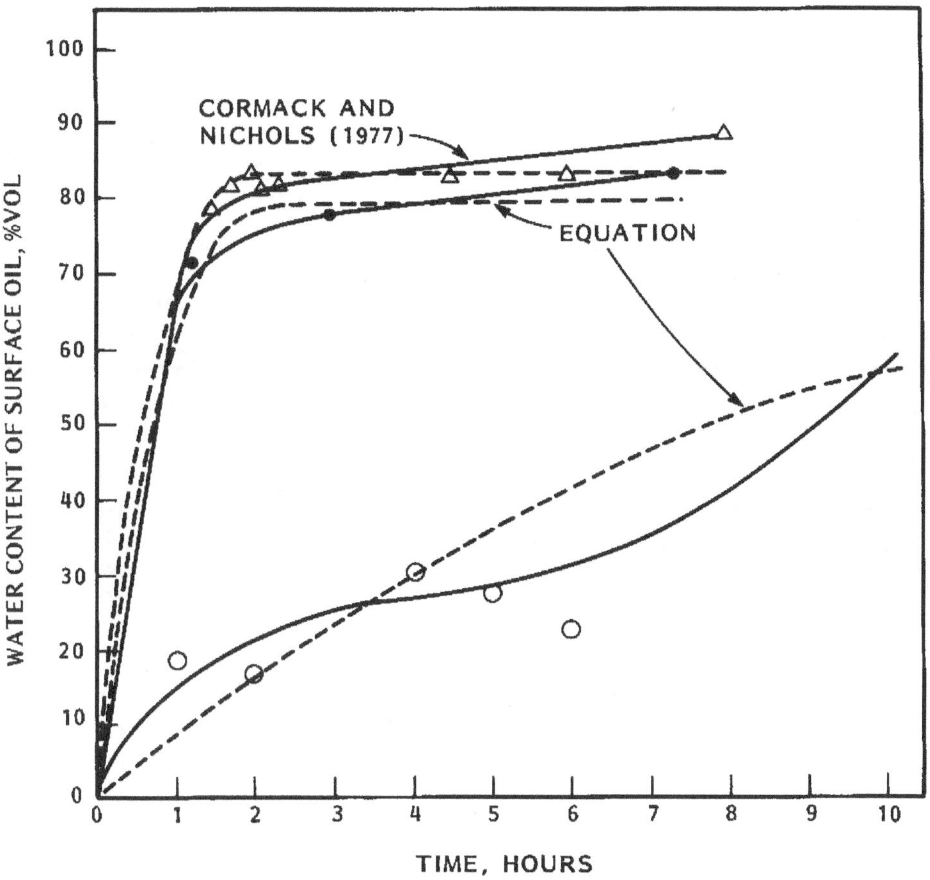

	WIND SPEED KNOTS	EQUATION CONSTANTS
● JULY 1975	12	K1 = 0.654 K2 = 1.266 K3 = 0.00110
○ JUNE 1976	2–3	K1 = 0.654 K2 = 1.230 K3 = 0.0000964
△ JAN 1977	15–30	K1 = 0.654 K2 = 1.194 K3 = 0.00130

Figure 12. COMPUTER MODEL - PREDICTED AND OBSERVED (CORMACK AND NICHOLS, 1977) TIME-DEPENDENT UPTAKE OF WATER INTO EKOFISK OIL AS A FUNCTION OF WIND SPEED AND SEA STATE. (FROM MACKAY et al., 1980.). REPRODUCED WITH PERMISSION OF THE AUTHOR.

and sea conditions, algorithms were also developed to describe the oil-in-water dispersion phenomena. Rapidly dispersing slicks will yield entirely different environmental impacts than emulsified slicks; thus, the capability of predicting which process would dominate would be invaluable in spill response countermeasures. For example, in situations where natural oil slick dispersion into the water column could be predicted, decisions concerning the use of chemical dispersants would be facilitated.

As noted in many of the case histories of major oil spills and in a number of laboratory studies, oil slicks exposed to sea surface turbulence tend to break up into oil droplets. Some of these droplets are sufficiently small that they are essentially incorporated into the water column where they later dissolve and are microbially degraded. The larger drops tend to return to the surface and presumably coalesce with the slick. The mechanisms for the formation of these droplets are far from clear, but Raj (1977) suggested that the dominant mechanism is wave breaking or white capping in which the water plunges into the oil slick, driving it under the surface. It is also possible that other mechanisms are operational when oil is dispersed under calmer, non-wave breaking conditions. Other factors such as rainfall, slick stretching-compression, interactions with sea-ice, or Langmuir circulations can carry oil into the water column.

Mackay et al. (1980) have examined the dispersion process on two laboratory scales, and they have developed theoretical equations which are reasonably consistent with experimental results. Oil is believed to be dispersed from the slick by the sum of two rates: R_B, a wave breaking mechanism; and R_N, a non-wave breaking mechanism. The units for both processes are $g/m^2 \cdot s$. The particulate oil droplets are initially dispersed into the well-mixed, near surface layer to a depth U (meters) which is determined by the depth of the plunging wave. While in this layer, the larger drops can rise and coalesce with the slick and, therefore, only temporarily dispersed. Smaller droplets with a lower rising velocity become permenently dispersed and diffuse (in part) vertically down into the less well-mixed layer of depth Z (meters), in which the vertical diffusivity (D) is m^2/s. With constant wave turbulence, the concentration of the small or permanently dispersed droplets is a constant C_S in the well-mixed layer and a variable C in the layer of depth Z. C increases with time at any given depth and concentrations are in g/m^3. The rate of diffusion into the less well-mixed layer, R_D, is the product of diffusivity, D, and the gradient in C at the interface between the layers where Z is equal to zero. That is,

$$R_D = D(dC/dZ)_{Z=0} = C_S K_1$$

where K_1 is the mass transfer coefficient (m/s).

Two different size classifications were considered for the model. Large drops (arbitrarily greater than 0.1 mm diameter) rise quickly at a velocity of V_L (m/s) and coalesce with the slick. Smaller drops (arbitrarily less than 0.1 mm diameter) have a lower rising velocity V_S (m/s) and may either coalesce or diffuse downwards. Their fate is thus controlled by these two competing processes. If the concentration of the large drops is C_L, then the rising flux must be $V_L C_L$ ($g/m^2 \cdot s$); similarily, for the small drops the rising flux is $V_S C_S$ ($g/m^2 \cdot s$). Adding these two fluxes to give the total coalescence rate R ($g/m^2 \cdot s$) yields:

$$R_C = V_L C_L + V_S C_S \ (g/m^2 \cdot s)$$

The differential equation for this upper layer is:

$$R_N + R_B - R_C - R_D = Ud(C_L + C_S)/dt$$

When a quasi-steady state is obtained the right hand side of the equation becomes zero.

If it is postulated that the volume fractions for the small dispersed oil drops for the two dispersion mechanisms are F_N and F_B for non-breaking and breaking sea states, respectively, the above equation can be broken down into small and large components. For small droplets:

$$F_N R_N + F_B R_B - V_S C_S - R_D = UdC_S/dt$$

Substituting $C_S K_1$ for R_D, separating variables, and integrating yields:

$$\ln(F_N R_N + F_B R_B - C_S[V_S + K_1]) = -t(V_S + K_1)/U + \text{constant}$$

Although the mass transfer coefficient, K_1, is time dependent, it was taken as a constant; at the initial condition, C_S was set to zero.

Thus,

$$C_S = \frac{(F_N R_N + F_B R_B)}{(V_S + K_1)} (1-\exp[-t(V_S + K_1)/U])$$

and at large time intervals, the quasi-steady state in C_S is reduced to:

$$C_S = (F_N R_N + F_B R_B)/(V_S + K_1)$$

Likewise, for large drops after extended time, the concentration can be shown to reduce to:

$$C_L = ([1-F_N]R_N + [1-F_B]R_B)/V_L$$

Details for these derivations are given in Mackay et al. (1980). Separate algorithms for non-breaking wave dispersion and breaking wave dispersion were derived. For the non-breaking wave dispersion mechanism it was suggested that the following equation applies:

$$R_N = K_2X^n$$

where R_N is the dispersion rate $(g/m^2 \cdot s)$, K_2 is a constant dependent on sea state, X is the slick thickness in meters, and n is a constant. This equation has the correct properties such that when X is zero, R_N is zero. Unfortunately, this equation also has the undesirable property that if n is negative, R_N approaches infinity when X goes to zero. Thus, Mackay et al. suggested that a better formulation where n can be positive would be:

$$R_N = K_A/(1+K_BX^n)$$

The dependence of K_2 on sea state in the first equation above can only be approximated; a reasonable form of dependence is:

$$K_2 = K_3S^m$$

where K_3 is a constant, S is the sea state, and m is a constant which was arbitrarily assigned a value of unity in absence of information to the contrary. From these equations it was possible to derive an algorithm for the slick life time τ:

$$\tau = X_o^{1.25}\rho/(1.25K_3S)$$

where X_o is the initial slick thickness and ρ is the oil density. For illustrative purposes, life times of slicks of variable thickness were then calculated. Illustrative data are presented below:

Initial Slick Thickness	τ -- Slick Lifetime
1 cm	290 days
1 mm	16 days
0.1 mm	22 hours
10 μm	75 minutes
1 μm	4.1 minutes

In summary then, it was suggested that for non-wave breaking conditions an expression for the dispersion rate into small droplets be used of the form

$$R_N F_N = K_3 S^m X^{-0.25} \quad (g/m^2 \cdot s)$$

where K_3 will probably have a value of approximately 10^{-4} and m a value of 1.

Similar derivations were presented for the wave-breaking dispersion model. However, the authors concluded that this mechanism, like the non-breaking dispersion mode, was only poorly understood and their equations could be criticized for containing an excessive number of adjustable parameters. Further, there were insufficient data to justify the equations complexities. Therefore, a simpler set of equations were derived and presented in finite difference form, suitable for incorporation into oil spill models. The breaking and non-breaking dispersion algorithms were then combined in one process; the rate of dispersion was given by:

$$F = K_A(U + 1)^2$$

where

F is the fraction of sea surface subject to dispersion per second
U is the wind speed
K_A is a constant, and

$$F_B = 1/(1 + K_B \mu^{0.5} X \cdot \gamma)$$

where

F_B is the fraction below the critical size
K_B is a constant
μ is viscosity
X is the slick thickness
γ is the oil-water interfacial tension.

The rate of dispersion is then, FX $(m^2/m^3 \cdot s)$, and the breakdown into large and small droplets was described as before.

With regard to modeling the thickness of oil and water-in-oil emulsions spreading at sea, Blokker (1964) has developed an equation for a circular oil spill where the slick diameter (D) in meters at a given time (t) can be determined as follows:

$$D^3 = \frac{24}{\Pi} K_r (d_w - d_o) \frac{d_o}{d_w} V_o t + D_o^3$$

where

> D is the slick diameter in meters
> d_w and d_o are the density values for water and oil, respectively (g/cm^3)
> V_o is the original oil volume, in meters^3
> t is the time following the oil spill, in minutes
> D_o is the immediate slick diameter at t = 0, and
> K_r is a constant, depending on the oil type.

From spreading experiments performed with various crude oils, Berridge et al. (1968a) modified the Blokker equation to estimate the relationship between slick thickness and the time following a spill. From this work the slick thickness was calculated as follows:

$$\text{thickness(cm)} = \frac{K^1}{t^{2/3}}$$

where:

> t is the time, in seconds.

$$K' = \left[\frac{v}{\pi}\right]^{1/3} \left[\frac{D_w}{3D_o(D_w - D_o)K_r}\right]^{2/3}$$

> v is the volume of oil, in cm^3
> D_o and D_w are the densities of oil and water, respectively (in g/cm^3)
> K_r is a constant, depending on the oil type.

Although neither of these models was designed to accommodate the emulsification process, Twardus (1980) stated that the change in spreading tendency caused by the water-in-oil emulsification could be approximated by substituting a higher value of oil density in both models. Thus, under the same conditions the equations predict that emulsified petroleum will spread less readily and will not spread as thinly over water as non-emulsified petroleum.

Grose (1979) also described a model that computes the thickness of bulk oil as a function of the physical characteristics of the oil and weathering parameters, determined by the formation of sheen and evaporation as influenced by wind speed. The model takes into account three generally observed phenomena:

1. Spilled oil does not form a single pool, but is generally composed of numerous patches of thick oil surrounded by thinner sheen;

2. The thickness of the patches is a function of the bulk physical properties of the oil and local environmental conditions including wave height, water temperature, and wind speed;

3. The weathering of the individual components are dependent upon the physical-chemical properties in the original oil.

The patch thickness of the bulk oil was estimated by balancing surface tension, buoyancy, and compressive forces after Grose (1978), which expanded upon the static equilibrium model derived by Langmuir (1933). The algorithm used is:

$$h = 0.01\cdot(C_d\cdot S_w\cdot W^2) + SQRT([C_d\cdot S_w\cdot W^2]^2 - 2\cdot FS\cdot E)/E$$

where h is the thickness in meters, C_d is the drag coefficient of the patch moving over the water surface, S_w is specific gravity of sea water, W is the relative speed of the patch through the surface water in cm/s, FS is the spreading force in dynes/cm, and E is the buoyancy factor:

$$E = 980\cdot S_o\cdot(S_w - S_o)/S_w$$

where S_o is the specific gravity of the bulk oil. The spreading forces range from +20 to -10, and are computed as the surface tension of water minus the sum of the surface tension of water and interfacial tensions of the oil.

The bulk density of the patch was computed by summing masses from each fraction, and dividing by the sum of the ratios of the masses to their characteristic densities (total volume). Evaporation is computed using the bulk properties of the fractions within the patch and the vapor pressures after Mackay and Matsugu (1973). The model does not include parameters for emulsification or incorporation of water which would change the specific density, viscosity, total surface area, and volume of the patches. Thus, some modifications would be required to predict the thickness of water-in-oil emulsions or mousse. This model was used to predict patch sizes of the Bunker C cargo lost from the Potomac off Greenland in 1977 (Petersen, 1978). The agreement between predicted and observed behavior was discussed in the case history of that spill.

Several models describing the break up and dispersion of oil slicks at sea have been presented (Mackay and Matsugu, 1973; Butler, 1975a; Grose, 1979; Mackay et al., 1979, 1980;

Aravamudan et al., 1981; Belen et al., 1981). With the exception of Butler (1975a), who used his model to predict the age of pelagic tar, most of these dispersion models do not deal directly with the ultimate weathering or fate of water-in-oil emulsions or mousse. The interested reader is, however, referred to the original articles for discussions of the oil weathering parameters.

With regard to the destabilization of water-in-oil emulsions, Twardus (1980) stated that the separation of water from a water-in-oil emulsion could be approximated by a Stokes law behavioral form in which the sedimentation (separation) rate (μ) of a spherical particle in a viscous liquid is given by the following equation:

$$\mu = 2gr^2 \frac{(d_1 - d_2)}{9\eta}$$

where:

- g equals the acceleration due to gravity
- r equals the particle radius
- d_1 equals the density of the particle
- d_2 equals the density of the liquid, and
- η equals the viscosity of the liquid

This equation shows that destabilization will be faster for water droplets with a larger radius and will be inversely proportional to the viscosity of the overall fluid (a prediction in line with observed stabilities of laboratory generated emulsions).

In many of these modeling efforts, it has been stated that complete mathematical descriptions of emulsion formation (and oil weathering in general) are presently impossible because of a lack of understanding of the system's basic physical-chemical properties (Mackay et al., 1980). Thus, although Grose (1979) achieved fair agreement between the predicted patch thickness and that observed in the field after the Potomac spill of Bunker C fuel oil in Melville Bay, he also stated that the model was limited by the lack of pertinent data to determine some of the transfer rates and coefficients. This was particularily true for further parameterization of the availability of each fractional component at the air/oil interface for evaporation and the parameterization for sheen generation at the oil/water interface, including in particular, the availability of material for generating sheen. The availability of material for generating sheen is constrained by the total mass in the fraction and the scale length related to the ratio of the thickness to the diameter of the patch. With the Availability of Material parameter in particular, the volumetric transfer rate into sheen was assumed to be dependent only on the inverse molecular weight of that fraction. The authors stated that this may not differentiate sufficiently between the more and less mobile fractions within the oil, and that it did not account for diffusion control of individual components as the viscosity of the material increased. Thus, as noted earlier, an arbitrary constant was imposed in the model to constrain the heavier fractions (>nC-16) from entering the sheen.

Obviously, much more work is required before model simulations will be able to predict accurately slick behavior and emulsification processes. Ultimately, it may be necessary to incorporate compound specific properties into such models to fully predict oil/mousse weathering for a variety of oils under different environmental conditions.

Research is currently under way in a number of laboratories to generate much of this needed compound specific data for algorithm development and model verification. Specifically, diffusivities of individual oil components for oils of different viscosities, and Henry's law constants for evaporation of specific components in different crude oils or petroleum product mixtures, need to be determined. Also, the detailed and specific effects of water-in-oil emulsification on these parameters has yet to be elucidated.

CHAPTER 6

SUMMARY, CONCLUSIONS, AND CRITICAL CITATION REVIEW

This chapter was designed as an overview of the topics considered in the rest of the text and to direct the reader to appropriate sections for additional information from which these conclusions were drawn. In preparing this literature review on water-in-oil emulsification processes and tar ball formation, every attempt was made to include as many original data as possible to illustrate the pertinent concepts discussed in each chapter. Obviously, it was impossible to include all such data from the rather voluminous amount of literature published on the subject. Therefore, the interested reader is referred to the original papers for additional details.

LABORATORY STUDIES

In laboratory studies to evaluate water-in-oil emulsion formation and behavior completed to date, it has been found that the results depend on the unique chemical compositions of the different crude oils and petroleum products tested. In general, heavier crudes with higher viscosities form the more stable emulsions (Bocard and Gatellier, 1981), and the presence of asphaltenes and higher molecular weight waxes is closely correlated with mousse stability (Berridge et al., 1968a, b; Davis and Gibbs, 1975; MacGregor and McLean, 1977; Mackay et al., 1979, 1980; Twardus, 1980; Bocard and Gatellier, 1981; Bridie et al., 1980a, b). Slightly different results have been obtained in other investigations, but it has generally been found that these materials act together in the emulsification process, although the asphaltenes appear to play a more important role (Bridie et al., 1980a, b; Berridge et al., 1968b). The crystallizing properties of the component waxes (near the pour points of the oils tested) are believed to be important in affecting the internal oil/mousse structure and viscosity; the asphaltenes are believed to act as surfactants which prevent water-water coalescence in the more stable mixtures (Berridge et al., 1968b; Canevari, 1969; Mackay et al., 1973; Bridie et al., 1980a, b; Cairns et al., 1974). Additional indigenous surface active agents such as metallo-porphyrins and other sulfur and oxygen compounds may be equally important. The products of photochemical and microbial oxidation have also been identified as having an important role as stabilizing agents. In several instances where these primary stabilizing components were not present, stable mousse could only be formed with photochemically or microbially weathered oils; notably this occurred with Brega, Nigerian, Zarzatine, and Lt. Arabian crudes (Berridge et al., 1968b; Bocard and Gatellier, 1981; Klein and Pilpel, 1974; Burwood and Spears, 1974; Zajic et al., 1974; Friede, 1973; Guire et al., 1973).

Stable mousses are not formed in laboratory studies at any temperatures with light petroleum distillates such as gasoline, kerosines, and several diesel fuels (Berridge et al., 1968b; Twardus, 1980). Interestingly, stable mousse formation could only be obtained in the laboratory with several light lube oils after they were fortified with wax and asphaltene mixtures obtained from known mousse forming oils such as Kuwait crude (Bridie et al., 1980a,b). This asphaltene mixture could also contain other high molecular weight surface active agents.

Temperature also affects mousse formation. In several instances (at lower temperatures approaching the pour point of the heavier oils), stable emulsions have been generated regardless of wax or asphaltene content. Conversely, if stable water-in-oil emulsions are repeatedly exposed to freeze-thaw cycles, some destabilization and separation of water and oil may occur (Dickens et al., 1981; Twardus, 1980). Similar results have been obtained when laboratory-generated and real spill water-in-oil emulsions are subjected to prolonged heating on removal from the water.

The absolute water content and sizes of water droplets incorporated into various mixtures of mousse also affect their stability and viscosity (Berridge et al., 1968b; Mackay et al., 1980; Twardus, 1980; Bocard and Gatellier, 1981). Positive correlations of percent water versus mousse stability and viscosity have been noted for several of the crude oils studied (Mackay et al., 1979, 1980). In general, the most stable emulsions from laboratory and field observations contain water droplets in the size range of from less than one to ten micrometers. Stable mousse can be formed with many oils containing from 20 to 80% water; however, above an oil-specific critical point, destabilization of the emulsions occurs (Berridge et al, 1968b; Twardus, 1980). Presumably, this reflects enhanced water-water contact and coalescence resulting in ultimate phase separation. Payne et al. (1984b) reported extremely rapid formation of a stable mousse in the presence of broken ice with turbulence in wave tank studies. Incorporation of water into the oil was promoted by the micro-scale turbulence caused by the grinding action of the ice floes and ice crystals. Rapid increases in percent water content in the emulsion were accompanied by an increased oil viscosity, without the corresponding changes in oil/water interfacial surface tension that characterize the mousse formation process under open-ocean, ice-free conditions (Payne et al., 1984a).

In most of the laboratory studies, the presence or absence of bacteria and suspended particulate material do not appear to affect emulsion behavior (Berridge et al., 1968a, b; Davis and Gibbs, 1975). Bacterial growth is generally limited to the surface of the mousse products tested and is believed to be inhibited by limited oxygen and nutrient diffusion into the mousse. Toxic materials inherent to the oils themselves may also be responsible for these observations, although water content and, in particular, the size of the water droplets encapsulated within the mixtures is correlated with the extent of bacterial infestation on the less stable emulsions (Berridge et al., 1968a, b). In several laboratory studies bacterial utilization of the mousse only occurred after treatment with dispersants, which resulted in break-up of the material with concomitant increases in surface-to-volume ratios (Bocard and Gatellier, 1981).

PHYSICAL PROPERTIES OF WATER-IN-OIL EMULSIONS

The physical properties of stable emulsions are impressively different from those of the starting crudes. Increases in specific gravity and viscosity have been observed to affect spreading, dispersion, and dissolution rates (Berridge et al., 1968b; Davis and Gibbs, 1975; MacGregor and McLean, 1977; Mackay et al., 1979, 1980; Twardus, 1980; Payne et al., 1981b, 1984a). Some evidence suggests that evaporation rates of intermediate molecular weight (C_9 to C_{12}) hydrocarbons from emulsions are affected (Payne et al., 1981b; Twardus, 1980; Nagata and Kondo, 1977). In general, the largest effects occur in the emulsions containing greater than 50% water. Water-in-oil emulsions with less water usually have pour points, spreading properties, and viscosities which proportionately resemble those of the starting oils (Twardus, 1980; Mackay et al., 1980).

The flash points and burn points of the water-in-oil emulsions studied also vary considerably with water content. For medium crudes, <u>in situ</u> combustion was inhibited when the water content reached 70%; for heavier crudes, inhibition of combustion occurred when the water content reached 30% (Twardus, 1980). Water dependent increases in viscosity also affect clean-up procedures of skimming and mopping, and pumping of these mixtures also becomes more difficult. The sorption capacity of various commercially available sorbant materials has also been observed to decrease as the water content in the mousse mixtures increases. This behavior may be related to the hydrophobic properties of the sorbant materials examined (Twardus, 1980).

TREATMENT OF MOUSSE WITH DISPERSANTS

Pre-treatment of oil and/or seawater with dispersants or demulsifiers generally inhibits laboratory mousse formation with most of the oils and petroleum products tested (Berridge et al., 1968b; Bridie et al., 1980a, b). In these studies, only 0.1 to 1% dispersant was required and, with several of the products tested, similar results were obtained when the dispersant was added to either the water or the oil. Previously stabilized mousse is much more difficult to break up with commercially available dispersants, although some success has been obtained with various products when sufficient mixing energy is utilized to thoroughly mix the dispersant into the water-in-oil mixture (Bridie et al., 1980b; Bocard and Gatellier, 1981; Lee et al., 1981). In general, however, no effective break-up of stabilized mousse could be achieved for water-in-oil emulsions with viscosities greater than 4,000 to 7,000 cP (Mackay et al., 1980; Lee et al., 1981).

The ineffectiveness of several of the dispersants studied to break up stable mousse formations at sea has been attributed to the lack of penetration of the dispersant into the mousse and its rapid removal from the mousse surface by waves and sea-surface turbulence (Lee et al., 1981). In several planned sea tests, mousse forming crudes such as La Rosa were effectively dispersed before mousse formation occurred (McAuliffe et al., 1981; JBF/API, 1976). Thicker lenses or patches of oil were observed to move along the leading (downwind) edge of these slicks, and dispersants were most effective when applied directly to the thicker lenses or patches rather than to the trailing sheen or thinner slick. Again, in the at-sea tests, mixed results were obtained depending on the type of dispersant/demulsifier used and the oil/mousse mixture tested. However, all dispersants work better when applied to the emulsions neat, rather than prediluted with seawater.

While many of the mousse formations have not been effectively broken up by demulsifier addition in laboratory tests, large and nearly immediate decreases in viscosities are often noted. In several clean-up operations the injection of demulsifier and dispersants into oil/mousse mixtures greatly enhanced pumping efficiency (Bridie et al., 1980b; Bocard and Gatellier, 1981).

CASE HISTORIES OF REAL SPILL EVENTS

In the case histories of major open-ocean and nearshore spills and blowouts, the formation and subsequent fates of the mousse were directly dependent upon the type of oil released. In general, oils which were capable of forming a stable water-in-oil emulsion or mousse in laboratory studies also formed mousse in the marine environment after real spill situations. The Kuwait crude oil cargoes from the Torrey Canyon (Smith, 1968) and Metula (Straughn, 1977; Hann, 1977) formed a very stable mousse, as predicted from results of the laboratory experiments. Likewise, the loss of the Arabian crude oil cargo and Bunker C fuel from the Amoco Cadiz resulted in the extremely rapid formation of a stable water-in-oil emulsion (Calder and Boehm, 1981; Hann et al., 1978). Weathering of Bunker C cargoes was somewhat more variable, with considerable mousse formation after the Arrow spill (Owens, 1978; Rashid, 1974; Mackay et al., 1973), but no apparent water-in-oil emulsification noted following the loss of the Bunker C cargo from the Potomac (Petersen, 1978). In this latter instance, however, the cargo contained 55% pitch, and the spill occurred in very calm seas. When the Arrow went aground, intense storm activities undoubtedly contributed to the 40 to 60% water-in-oil emulsion formation. In laboratory studies using Bunker C, a rigid and sticky water-in-oil emulsion forms with water contents up to 60% (Berridge et al., 1968b). In line with previous laboratory and field experiments using similar oils, no "mousse" formation was observed after the Burmah Agate spill, where a light Nigerian Crude (39.3° API) was released (Kelley et al., 1981). Similarly, no mousse formation occurred following an actual spill of JP-5 (aircraft fuel) from the M/V Cepheus near Anchorage, Alaska (Payne and McNabb, unpublished data).

In both the Arrow and Potomac spill incidents, break-up of the cargo into sub-micrometer to several centimeter sized oil/mousse particulates was observed. In the Arrow spill break-up of the oil occurred at relatively large distances from the spill site (Forrester, 1971). Ingestion of these micro-particulates by pelagic organisms was noted, and incorporation of oil into fecal pellets was postulated as a mechanism for removing such droplets from the water column (Conover, 1971).

In the two open ocean blowouts considered in this review, markedly different results were observed. In the Ekofisk Bravo blowout, a very unstable yellowish to brown water-in-oil emulsion (35 to 72% water) formed, but this material rapidly broke up into smaller particles and tar patches and was transported from the wellhead with little long-term environmental damage (Cormack and Nichols, 1977; Grahl-Nielson, 1977; Audunson, 1978). A slightly more stable mousse formed after increased sea-surface weathering, but this too was unstable and disappeared after several weeks. It should be noted, however, that this crude contains very low levels of stabilizing agents (asphaltenes 0.03%). After the IXTOC I blowout in the Bay of Campeche, Gulf of Mexico, mousse formation occurred only after appreciable evaporation, dissolution, and photochemical weathering of the oil (Atwood et al., 1980; Overton et al., 1980a). The mousse resulting from these processes was extremely stable, and extensive areas of much of the Gulf of Mexico and its adjacent shorelines subsequently were covered and contaminated with mousse and tar.

The extent of microbial degradation of the oils released from the spills and blowouts considered was extremely variable. In the Amoco Cadiz spill off Brittany, France, microbial processes were observed to compete very effectively with evaporation and dissolution weathering for removal of specific components in the oil (Calder and Boehm, 1981). In the IXTOC I blowout and the Potomac spill, on the other hand, little, if any, microbial degradation occurred to the oil or mousse in the water column over extended periods of time (IXTOC--Boehm and Fiest, 1980a; Overton et al., 1980b; Payne et al., 1980a; Buckley et al., 1980; Atlas et al., 1980; Potomac--Petersen, 1978). Long-term microbial processes were important in the fate of beached and stranded oil and mousse in the Torrey Canyon (Smith, 1968), Arrow (Vandermeulen et al., 1977), Metula (Straughn, 1977), and Amoco Cadiz (Calder and Boehm, 1981; Boehm et al., 1981; Ward, 1981; Vandermeulen et al., 1981) spills, whereas little evidence of microbial degradation of stranded IXTOC-I mousse was observed after several weeks in samples collected up to 500 miles from the wellhead along the upper intertidal shores of Laguna Madre (Payne et al., 1980a; Overton et al., 1980b).

The long term environmental impact associated with the oil and mousse that reaches shorelines appears to be directly correlated with the amount of coverage, the intertidal substrate type, and the energy regime at the site of the oil/mousse stranding (Blumer et al., 1973). Without exception, the most deleterious effects have been observed in low energy lagoons, estuaries, and bays, where stranded oil may persist for upwards of 10 years.

TAR BALL DISTRIBUTIONS AND CHEMISTRY

Tar ball concentrations and compositions vary throughout the world's oceans, but localized "hot spots" of tar ball accumulations have been reported in several areas. Quite often these higher concentrations occur along known lines of tanker traffic or in currents downstream from these areas. In general, the Mediterranean Sea and the Sargasso Sea have similar high levels of tar balls which, in turn, are greater than levels observed in the broader expanses of the North Atlantic and North Pacific oceans.

Chemical compositions of tar balls are extremely variable, and the physical appearance can range from brittle and hard to soft and very sticky, with sizes ranging from several mm to tens of cm (Butler, 1973; Koons, 1973; Mommessin and Raia, 1975; Wade et al., 1976; Jeffrey et al., 1974; Jordan and Payne, 1980). In general, compounds with molecular weights less than n-C_{15} are not present, and most of the tar balls considered have very high residues of compounds with molecular weights greater than n-C_{34}. Several sources for these tar balls have been identified, including: tanker cargo-hold washings, bilge discharges, urban/industrial wastes, and residues from major spills or blowouts such as the Amoco Cadiz and IXTOC I. In some instances the chemical composition of separate tar balls collected from localized areas are very similar; however, remarkably different chemical compositions of separate tar balls collected in the same neuston tows have been noted (Mommesin and Raia, 1975). Because of the extreme patchiness of tar balls, and the several orders of magnitude variation in tar ball loadings in the same area over the period of a single day, standing stock estimates of tar ball pollution are tenuous at best. Therefore, it is not possible at this time to determine if tar ball occurrence is increasing or decreasing.

The ultimate fate of most of these tar balls at sea probably consists of break-up and sinking within a period of one year (Morris, 1971; Butler et al., 1973; Butler, 1975a, b; Horn et al., 1970; Brown et al., 1973; Brown et al., 1975; Brown and Huffman, 1976; Wade et al., 1976). In contrast, beached or stranded tar balls have a fate similar to that of larger patches of mousse or oil released from major spill incidences. Tar ball sitings on beaches have been reported globally; however, most evidence suggests that their levels probably have not changed over the last decade. The decomposition of stranded tar balls is extremely dependent on the shoreline substrate, the energy regime of the shoreline environment, the presence or absence of nutrients, and the rate of sedimentation within the intertidal zone (Blumer et al., 1973). Microbial degradation of tar balls generally is limited to the outer surfaces due to diffusion limited transport of oxygen and nutrients to the interior. The fact that numerous tar balls do not have a detectable component concentration gradient with depth into the interior, suggests that evaporation, dissolution, and photochemical and microbial degradation occurs before agglomeration and tar ball formation.

MATHEMATICAL AND COMPUTER MODELING OF MOUSSE BEHAVIOR

In reviewing the status of mathematical and computer modeling of mousse formation and degradation, it was noted that numerous mathematical formulations have been developed to describe individual aspects of water-in-oil emulsion formation and behavior (Mackay et al., 1979, 1980; Twardus, 1980; Raj, 1977; Grose, 1979). Unfortunately, however, no single oil weathering model presently exists that accounts for all of the variable factors. It is impossible to completely model water-in-oil emulsion formation and behavior because of the lack of a sound understanding of all of the physical and chemical properties of the system. Considerable advances have been made in modeling viscosity changes of water-in-oil emulsions as a function of water content (Mackay et al., 1979, 1980), the competitive process of dispersion of oil and mousse into the water column (Mackay et al., 1979, 1980), and the break-up or decomposition of slicks or patches of oil based on evaporation and dissolution weathering (Mackay and Matsugu, 1973; Butler, 1975a; Grose, 1979; Mackay et al., 1979, 1980; Aravamudan et al., 1981; Belen et al., 1981; Payne et al., 1981b, 1984a,b). Nevertheless, much more work will be required before model simulations will be able to predict accurately slick behavior and emulsification processes. Ultimately, it may be necessary to incorporate compound-specific properties into modeling efforts to fully predict oil/mousse/tar ball weathering for a variety of oils under different environmental regimes (Payne et al., 1981b, 1984a,b).

BIBLIOGRAPHY

Addy, J. M., D. Levell, and J. P. Hartley. 1978. Biological Monitoring of Sediments in Ekofisk Oilfield. Pages 514-519 In Proceedings, Conference on Assessment of Ecological Impacts of Oil Spills (June 14-17, Keystone, Colo.). American Institute of Biological Sciences. Washington, D.C.

Aminot, A. 1981. Anomalies de Systeme Hydrobiologique Cotier Apres L'echouge de L'Amoco Cadiz. Considerations Qualitatives et Quantitatives sur la Biodegradation in Situ des Hydrocarbures. Pages 223-242 In, Amoco Cadiz Fates and Effects of the Oil Spill. Proceedings of the International Symposium Centre Oceanologique de Bretogne, Brest (France).

Aravamudan, K. S., P. K. Raj, and Lt. G. Marsh. 1981. Simplified Models to Predict the Breakup of Oil on Rough Seas. Pages 153-159 In Proceedings, 1981 Oil Spill Conference. American Petroleum Institute, Washington, D.C.

Atlas, R. M., G. Roubal, A. Bronner, and J. Haines. 1980. Microbial Degradation of Hydrocarbons in Mousse from IXTOC-I. Pages 411-435 In Proceedings of a Symposium, Preliminary Results from the September 1979 Researcher/Pierce IXTOC-I Cruise (June 9-10, Key Biscayne, Fla.). NOAA, Office of Marine Pollution Assessment. Washington, D.C.

Attaway, D., J. R. Jadamec, and W. McGowan. 1973. Rust in Floating Petroleum Found in the Marine Environment. Unpublished Manuscript, U.S. Coast Guard. Cited in: Figs. 7 & 8 of J. N. Butler, B. F. Morris, and J. Dass, 1973. Pelagic Tar from Bermuda and the Sargasso Sea. Bermuda Biol. Station Spec. Publ. #10, pp. 23.

Atwood, D. K., J. A. Benjamin, and J. W. Farrington. 1980. The Mission of the September 1979 Researcher/Pierce IXTOC-I Cruise and the Physical Situation Encountered. Pages 1-16 In Proceedings of a Symposium, Preliminary Results from the September 1979 Researcher/Pierce IXTOC-I Cruise (June 9-10, Key Biscayne, Fla.). NOAA, Office of Marine Pollution Assessment. Washington, D.C.

Audunson, T. 1978. The Fate and Weathering of Surface Oil from the Bravo Blowout. Pages 445-475 In Proceedings, Conference on Assessment of Ecological Impacts of Oil Spills (June 14-17, Keystone, Colo.). American Institute of Biological Sciences. Washington, D.C.

Balkas, T. I., I. Salihoglu, A. F. Gaines, M. Sunay, and J. Matthews. 1982. Characterization of Floating and Sinking Tar Balls in the Marine Environment. Mar. Poll. Bull. 13:202-205.

Barber, F. G. 1970. Report of the Task Force: Operation Oil (Cleanup of Arrow Oil Spill in Chedabucto Bay). Pages 35-54. In Ministry of Transport, Ottawa, Volume 3.

Becher, P. 1955. Principles of Emulsion Technology. Reinhold Publishing Corp. New York, N.Y.

Bedinger, C. A., Jr. and C. P. Nulton. 1982. Analysis of Environmental and Tar Samples from the Nearshore South Texas Area After Oiling from the IXTOC-I Blowout. Bull. Environm. Contam. Toxicol. 28:166-171.

Belen, M. S., W. J. Lechr, and H. M. Cekirge. 1981. Spreading, Dispersion, and Evaporation of Oil Slicks in the Arabian Gulf. Pages 161-164 In Proceedings, 1981 Oil Spill Conference. American Petroleum Institute, Washington, D.C.

Benzhitskiy, A. G. 1981. Distribution of Tar Balls in the Surface Layer of the Arabian Sea in April-June 1980. Oceanology 21:717-720.

Berridge, S. A., R. A. Dean, R. G. Fallows, and A. Fish. 1968a. The Properties of Persistent Oils at Sea. Pages 2-11 In P. Hepple, ed. In Proceedings of a Symposium, Scientific Aspects of Pollution of the Sea by Oil.

Berridge, S. A., M. T. Thew, and A. G. Loriston-Clarke. 1968b. The Formation and Stability of Emulsions of Water in Crude Petroleum and Similar Rocks. Journal of the Institute of Petroleum. 54:333-357.

Blokker, P. C. 1964. Spreading and Evaporation of Petroleum Products on Water. Presented at the 4th International Harbour Conference. June 22-27, 1964 Antwerp, Belgium.

Blumer, M., M. Ehrhardt, and J. H. Jones. 1973. The Environmental Fate of Stranded Crude Oil. Deep Sea Res. 20:239-259.

Bocard, C. and C. Gatellier. 1981. Breaking of Fresh and Weathered Emulsions by Chemicals. Pages 601-607 In Proceedings, 1981 Oil Spill Conference. American Petroleum Institute, Washington, D.C.

Boehm, P. D. and D. L. Fiest. 1980a. Surface Water Column Transport and Weathering of Petroleum Hydrocarbons During the IXTOC-I Blowout in the Bay of Campeche and Their Relation to Surface Oil and Microlayer Compositions. Pages 267-338 In Proceedings of a Symposium, Preliminary Results from the September 1979 Researcher/Pierce IXTOC-I Cruise. (June 9-10, Key Biscayne, FL) NOAA, Office of Marine Pollution Assessment. Washington, D.C.

Boehm, P. D. and D. L. Fiest. 1980b. Aspects of the Transport of Petroleum Hydrocarbons to the Offshore Benthos During the IXTOC-I Blowout in the Bay of Campeche. Pages 207-236 In Proceedings of a Symposium, Preliminary Results from the September 1979 Researcher/Pierce IXTOC-I Cruise. (June 9-10, Key Biscayne, FL) NOAA, Office of Marine Pollution Assessment. Washington, D.C.

Boehm, P. D., D. L. Fiest, and A. Elskus. 1981. Comparative Weathering Patterns of Hydrocarbons from the Amoco Cadiz Oil Spill Observed at a Variety of Coastal Environments. Pages 159-173 In Proceedings of the International Symposium, Amoco Cadiz Fates and Effects of the Oil Spill. Centre Oceanologique de Bretagne, Brest (France).

Bridie, A. L., Th. H. Wanders, W. Zegveld, and H. B. Van der Heijde. 1980a. The Formation, Prevention and Breaking of Sea-Water-in-Crude-Oil Emulsions: "Chocolate Mousse". Pages 33-39 In International Research Symposium, Chemical Dispersion of Oil Spills (Nov. 17-19, Toronto, Canada).

Bridie, A. L., Th. H. Wanders, W. Zegveld, and H. B. Van der Heijde. 1980b. Formation, Prevention and Breaking of Sea Water in Crude Oil Emulsions "Chocolate Mousses". Mar. Poll. Bull. 2:343-348.

Brooks, J. M., D. A. Weisenburg, R. A. Burke, M. C. Kenicutt, and B. B. Bernard. 1980. Gaseous and Volatile Hydrocarbons in the Gulf of Mexico Following the IXTOC-I Blowout. Pages 53-85 In Proceedings of a Symposium, Preliminary Results from the September 1979 Researcher/Pierce IXTOC-I Cruise. (June 9-10, Key Biscayne FL). NOAA, Office of Marine Pollution Assessment. Washington, D.C.

Brown, R. A., T. D. Searl, J. J. Elliot, B. G. Phillips, D. E. Brandon, and P. H. Monaghan. 1973. Distribution of Heavy Hydrocarbons in Some Atlantic Ocean Waters. Pages 505-519 In Proceedings, 1973 Joint Conference on Prevention and Control of Oil Spills. American Petroleum Inst., Washington, D. C.

Brown, R. A., J. J. Elliot, J. M. Kelliher, and T. D. Searl. 1975. Sampling and Analysis of Nonvolatile Hydrocarbons in Ocean Water. Pages 172-187 In Analytical Methods in Oceanography. Am. Chem. Soc. Adv. Ser. 147.

Brown, R. A. and H. L. Huffman, Jr. 1976. Hydrocarbons in Open Ocean Waters. Science 191:847-849.

Buckley, E. N., F. K. Pfaender, and K. L. Kylberg. 1980. Response of the Pelagic Community to Oil from the IXTOC-I Blowout: II. Model Ecosystem Studies. Pages 563-586 In Proceedings of a Symposium, Preliminary Results from the September 1979 Researcher/Pierce IXTOC-I Cruise. (June 9-10, Key Biscayne, FL) NOAA, Office of Marine Pollution Assessment. Washington, D.C.

Burns, K. A., J. P. Villeneuve, V. C. Anderlini, and S. W. Fowler. 1982. Survey of Tar, Hydrocarbon and Metal Pollution in the Coastal Waters of Oman. Mar. Poll. Bull. 13:240-247.

Burwood, R. and G. C. Spears. 1974. Photo-oxidation as a Factor in the Environmental Dispersal of Crude Oil. Estuarine Coastal Marine Science 2:117-135.

Butler, J. N. 1973. The Occurrence and Amount of Pelagic Tar in the Open Ocean. Pages 376-399 In Background Papers for a Workshop on Inputs, Fates, and Effects of Petroleum in the Marine Environment, Volume II. Prepared under the Aegis of Ocean Affairs Board, National Academy of Sciences, Washington, D.C.

Butler, J. N. 1975a. Evaporative Weathering of Petroleum Residues: the Age of Pelagic Tar. Marine Chemistry, 3:9-21.

Butler, J. N. 1975b. Pelagic Tar. Scientific America 232:90-97.

Butler, J. N., B. F. Morris, and J. Sass. 1973. Pelagic Tar from Bermuda and the Sargasso Sea. Special Publication No. 10, Bermuda Biological Station. 346 pp.

Butler, J. N., B. F. Morris, and T. D. Sleeter. 1976. The Fate of Petroleum in the Open Ocean. Pages 287-297 In Proceedings of the Symposium, Sources, Effects and Sinks of Hydrocarbons in the Aquatic Environment. The American Institute of Biological Sciences, (August 9-11, 1976), Washington, D.C..

Cairns, R. J. R., D. M. Grist, and E. L. Neustadter. 1974. The Effect of Crude Oil-Water Interfacial Properties on Water-Crude Oil Emulsion Stability. Pages 135-151 In A. L. Smith, ed. Theory and Practice of Emulsion Technology. Academic Press, New York.

Calder, J. A., J. Lake, and J. Laseter. 1978. Chemical Composition of Selected Environmental and Petroleum Samples from the Amoco Cadiz Oil Spill. Pages 21-84 In the Amoco Cadiz Oil Spill. A NOAA/EPA Special Report. NOAA, Washington, D.C.

Calder, J. A., and P. D. Boehm. 1981. The Chemistry of Amoco Cadiz Oil in Aber Wrac'h. Pages 149-158 In Proceedings of the International Symposium, Amoco Cadiz Fates and Effects of the Oil Spill. Centre Oceanologique de Bretagne, Brest (France).

Callahan, R. A. and A. Soutar. 1976. Southern California Baseline Studies - Collection of Analytical Samples. In Proceedings, 2nd Annual Conference Marine Technology Soc. Monthly News 29:667-674.

Canevari, G. P. 1969. General Dispersant Theory. Pages 171-177 In Proceedings, Joint Conference on Prevention and Control of Oil Spills. NTIS Report No. PB 194 395.

Canevari, G. P. 1982. The Formulation of an Effective Demulsifier for Oil Spill Emulsions. Mar. Poll. Bull. 13:49-54.

Clark, R. C., Jr. and W. D. MacLeod, Jr. 1977. Inputs, Transport Mechanisms, and Observed Concentrations of Petroleum in the Marine Environment. Pages 91-223 In D. C. Malins, ed. Effects of Petroleum on Arctic and Subarctic Marine Environments and Organisms, Volume I. Nature and Fate of Petroleum. Academic Press, Inc. New York, N.Y.

Coleman, H. J., E. M. Shelton, D. T. Nichols, and C. J. Thompson. 1978. Analyses of 800 Crude Oils from the United States Oil Fields. BETC/RI-78/14, Bartlesville Energy Technology Center, Bartlesville, OK.

Conomos, T. J. 1975. Movement of Spilled Oil as Predicted by Estuarine Nontidal Drift. Limnol. Oceanogr. 20:159-173.

Conover, R. J. 1971. Some Relationships Between Zooplankton and Bunker C Oil in Chedabucto Bay Following the Wreck of the Tanker Arrow. Journal of the Fisheries Research Board of Canada 28:1327-1330.

Cordes, C., L. Atkinson, R. Lee, and J. Blanton. 1980. Pelagic Tar off Georgia and Florida in Relation to Physical Processes. Mar. Poll. Bull. 11:315-317.

Cormack, D. and J. A. Nichols. 1977. The Concentrations of Oil in Seawater Resulting from Natural and Chemically Induced Dispersion of Oil Slicks. Pages 381-385 In Proceedings, 1977 Oil Spill Conference American Petroleum Institute, Washington, D.C.

Corredor, J. E., J. Morell, and A. Mendez. 1983. Pelagic Petroleum Pollution off the South-west Coast of Puerto Rico. Mar. Poll. Bull. 14:166-168.

Davis, S. J. and C. F. Gibbs. 1975. The Effect of Weathering on a Crude Oil Exposed at Sea. Water Research 9:275-289.

Dickens, D. F., I. A. Buist, and W. M. Pistruzak. 1981. Dome's Petroleum Study of Oil and Gas Under Sea Ice. Pages 183-189 In Proceedings, 1981 Oil Spill Conference American Petroleum Institute, Washington, D.C.

Dwivedi, S. N. and A. H. Parulekar. 1974. Oil Pollution Along the Indian Coastline. Marine Poll. Monitoring (Petroleum). Nat. Bur. Stand. Spec. Pub. No. 409:101-105.

Farrington, J. W. and B. W. Tripp. 1975. A Comparison of Analysis Methods for Hydrocarbons in Surface Sediments. Pages 267-284 In T. M. Church, ed. Marine Chemistry in the Coastal Environment. ACS Symposium Series No. 18. American Chemical Society. Washington, D.C.

Feist, D. L. and P. D. Boehm. 1980. Subsurface Distributions of Petroleum from an Offshore Well Blowout, Bay of Campeche. Pages 169-185. In Proceedings of a Symposium on Preliminary Results from the September 1979 Researcher/Pierce IXTOC-I Cruise (June 9-10, 1980, Key Biscayne, FL) NOAA, Office of Marine Pollution Assessment, Washington, D.C.

Feldman, M. H. 1973. Some Mechanisms of Weathering of Petroleum Hydrocarbons on Marine Waters: Competitive Pathways to Fate and Disposition of Petroleum Pollution. Pages 431-445 In Background Papers for a Workshop on Inputs, Fates, and Effects of Petroleum in the Marine Environment, Volume II. Prepared under the Aegis of

Ocean Affairs Board, National Academy of Sciences, Washington, D.C.

Forrester, W. D. 1971. Distribution of Suspended Oil Particles Following the Grounding of the Tanker Arrow. Journal of Marine Research 29:151-170.

Frankenfeld, J. W. 1973. Weathering of Oil at Sea. Submitted to the U.S. Coast Guard, Washington, D.C. NTIS Report No. AD 787 789.

Friede, J. D. 1973. The Isolation and Chemical and Biological Properties of Microbial and Emulsifying Agents for Hydrocarbons. Progress Report. AD 770-630. National Technical Information Service, U.S. Dept. of Commerce, Springfield, VA., 5 pp.

Geyer, R. A. 1981. Naturally Occurring Hydrocarbons in the Gulf of Mexico and the Caribbean. Pages 445-451 In Proceedings, 1981 Oil Spill Conference American Petroleum Institute, Washington, D.C.

Golik, A. 1982. The Distribution and Behavior of Tar Balls Along the Israeli Coast. Estuarine, Coastal and Shelf Science 15:267-276.

Gorden, D. C., Jr., P. D. Keizer, and N. J. Prouse. 1973. Laboratory Studies on the Accommodation of Some Crude and Residual Fuel Oils in Seawater. Journal of Fisheries Research Board of Canada 30:1611-1618.

Grahl-Nielsen, O. 1978. The Ekofisk Bravo Blowout. Petroleum Hydrocarbons in the Sea. Pages 477-499 In Proceedings of the Conference on Assessment of Ecological Impacts of Oil Spills. (June 14-17, Keystone, CO) American Institute of Biological Sciences. Washington, D.C.

Griffiths, R. P. and R. Y. Morita. 1981. Study of Microbial Activity and Crude Oil - Microbial Interactions in the Water and Sediments of Cook Inlet and the Beaufort Sea. Final Report RU 190, Submitted to NOAA/OCSEAP, Juneau, Alaska. 367 pp.

Grose, P. L. and J. S. Mattson. 1977. The Argo Merchant Oil Spill: A Preliminary Scientific Report. A NOAA/EPA Special Report, Washington, D.C. 133 pp.

Grose, P. L. 1978. The Behavior of Floating Oil from the Argo Merchant. In the Wake of the Argo Merchant. Pages 19-21 In Proceedings of a Symposium at The Center for Ocean Management Studies, University of Rhode Island.

Grose, P. L. 1979. A Preliminary Model to Predict the Thickness Distribution of Spilled Oil. Pages 1.1-1.16 In Workshop on the Physical Behavior of Oil in the Marine Environment. (May 8-9, Princeton Univ.) Prepared for National Weather Service, Silver Spring, MD.

Guire, P. E., J. D. Friede, and R. K. Gholson. 1973. Production and Characterization of Emulsifying Factors from Hydrocarbonoclastic Yeast and Bacteria. Pages 229-231 In D. G. Ahern and S. P. Meyers, eds. The Microbial Degradation of Oil Pollutants. Publ. No. LSU-SG-73-01. Center for Wetland Resources, LSU, Baton Rouge, LA.

Gundlach, E. R., P. D. Boehm, M. Marchand, R. M. Atlas, D. M. Ward, and D. A. Wolfe. 1983. The Fate of Amoco Cadiz Oil. Science 221:122-129.

Gundlach, E. R., K. J. Finkelstein, and J. L. Sadd. 1981. Impact and Persistence of IXTOC-I Oil on the South Texas Coast. Pages 477-485 In Proceedings, 1981 Oil Spill Conference. American Petroleum Institute, Washington, D.C.

Hann, R. W., Jr. 1977. Fate of Oil from the Supertanker Metula. Pages 465-473 In Proceedings, 1977 Oil Spill Conference. American Petroleum Institute, Washington, D.C.

Hayes, M. O., E. R. Gundlach, and L. D'Ozouville. 1979. Role of Dynamic Coastal Processes in the Impact and Dispersal of the Amoco Cadiz Oil Spill (March, 1978) Brittany, France. Pages 193-198 In Proceedings, 1979 Oil Spill Conference American Petroleum Institute. Washington, D.C.

Hollinger, J. P. and R. A. Menella. 1973. Oil Spills: Measurements of the Distributions and Volumes by Multi-Frequency Microwave Radiometry. Science 181:54-56.

Honjo, S. 1980. Material Fluxes and Modes of Sedimentation in the Mesopelagic and Bathypelagic Zones. J. Mar. Res. 38:53-97.

Horn, M. H., J. M. Teal, and R. H. Backus. 1970. Petroleum Lumps on the Surface of the Sea. Science 168:245-246.

JBF/API. 1976. Physical and Chemical Behavior of Crude Oil Slicks on the Ocean. API Publication 4290, American Petroleum Institute, Washington, D.C.

Jeffrey, L. M. 1973. Preliminary Report on Floating Tar Balls in the Gulf of Mexico and Caribbean Sea. Unpublished report. Sea Grant Project 53399, Texas A&M U., College Station, TX.

Jeffrey, L. M., D. J. Frank, N. Powell, A. Bautz, A. Vos, and L. May. 1973. Progress Report on Pelagic, Beach and Bottom Tars of the Gulf of Mexico and Controlled Weathering Experiments. Department of Oceanography, Texas A&M University, 86 pp.

Jeffrey, L. M., W. E. Pequegnat, E. A. Kennedy, A. Vos, and B. M. James. 1974. Pelagic Tar in the Gulf of Mexico and Caribbean Sea. Mar. Poll. Monitoring (Petroleum). Natl. Bur. Stand. Spec. Pub. 409:233-235.

Johnson, J. H., P. W. Brooks, A. K. Aldridge, and S. J. Rowland. 1978. Presence and Sources of Oil in the Sediment and Benthic Community Surrounding the Ekofisk Field After the Blowout at Bravo. Pages 488-513 _In_ Proceedings, Conference on Assessment of Ecological Impacts of Oil Spills. (June 14-17, Keystone, CO) American Institute of Biological Sciences. Washington, D.C.

Jordan, R. E. and Payne, J. R. 1980. Fate and Weathering of Petroleum Spills in the Marine Environment, Ann Arbor Science, Ann Arbor, Michigan. 174 pp.

Keizer, P. D., T. P. Ahern, J. Dale, and J. H. Vandermeulen. 1978. Residues of Bunker C Oil in Chedabucto Bay, Nova Scotia, 6 Years After the _Arrow_ Spill. Journal of the Fisheries Research Board of Canada 35:528-535.

Kelley, F. J., Jr., R. W. Hann, Jr., and H. N. Young. 1981. Surveillance, Tracking, and Model Correlation of the Spill from the Tanker _Burmah Agate_. Pages 147-152 _In_ Proceedings, 1981 Oil Spill Conference. American Petroleum Institute. Washington, D.C.

Klein, A. E. and N. Pilpel. 1974. The Effects of Artificial Sunlight Upon Floating Oils. Water Research 8:79-83.

Knap, A. H., T. M. Iliffe, and J. N. Butler. 1980. Has the Amount of Tar on the Open Ocean Changed in the Past Decade? Mar. Poll. Bull. 11:161-164.

Kolpack, R. L., R. W. Stearns, and G. L. Armstrong. 1978. Sinking of Oil in Los Angeles Harbor, California, Following the Destruction of the _Sansinena_. Pages 379-392 _In_ Proceedings, Conference on Assessment of Ecological

Impacts of Oil Spills. American Institute of Biological Sciences. Washington, D.C.

Koons, C. B. 1973. Chemical Composition: A Control on the Physical and Chemical Processes Acting on Petroleum in the Marine Environment. Pages 475-484 In Background Papers for a Workshop on Inputs, Fates, and Effects of Petroleum in the Marine Environment, Volume II. Prepared under the Aegis of Ocean Affairs Board, National Academy of Sciences, Washington, D.C.

Langmuir, I. 1933. Oil Lenses on Water and the Nature of Monomolecular Expanded Films. J. of Chem. Physics V1:756-776.

Lee, M., F. Martinelli, B. Lynch, and P. R. Morris. 1981. The Use of Dispersants on Viscous Fuel Oils and Water in Crude Oil Emulsions. Pages 31-35 In Proceedings, 1981 Oil Spill Conference American Petroleum Institute. Washington, D.C.

Little, R. C. 1981. Chemical Demulsification of Aged, Crude Oil Emulsions. Environ. Sci. Tech. 15:1184-1190.

Ludwig, H. F. and R. Carter. 1961. Analytical Characteristics of Oil-Tar Materials on Southern California Beaches. J. Wat. Poll. Control Fed. 33:1123-1139.

Macauley, M. C., K. Daly, and T. Saunders English. 1980. Acoustic Observations of Biological Volume Scattering in the Vicinity of the IXTOC-I Blowout. Pages 499-521 In Proceedings of a Symposium, Preliminary Results from the September 1979 Researcher/Pierce IXTOC-I Cruise. (June 9-10, Key Biscayne, FL) NOAA, Office of Marine Pollution Assessment. Washington, D.C.

MacGregor, C. and A. Y. McLean. 1977. Fate of Crude Oil Spilled in a Simulated Arctic Environment. Pages 461-463 In Proceedings, 1977 Oil Spill Conference, American Petroleum Institute. Washington, D.C.

Mackay, G. C. M., A. Y. McLean, O. J. Betancourt, and B. C. Johnson. 1973. The Formation of Water-in-Oil Emulsions Subsequent to an Oil Spill. Journal of the Institute of Petroleum 59:164-172.

Mackay, D. and R. S. Matsugu. 1973. Evaporation Rates of Liquid Hydrocarbon Spills on Land and Water. Canadian J. Chem. Eng. 51:434-439.

Mackay, D., I. Buist, R. Mascarenhas, and S. Patterson. 1979. Experimental Studies of Dispersion and Emulsion Formation from Oil Slicks. Pages 1.17-1.40 In Workshop on the Physical Behavior of Oil in the Marine Environment. Princeton Univ., prepared for the National Weather Service, Silver Spring, MD.

Mackay, D., I. Buist, R. Mascarenhas, and S. Patterson. 1980. Oil Spill Processes and Models. A Report Submitted to Environmental Emergency Branch Environmental Impact Control Directorate Environment Protection Service Environment Canada, (December) Ottowa, Ontario K1A 1C8.

Mao, M. L. M. and S. S. Marsden. 1977. Stability of Crude Oil-in-Water Emulsions as a Function of Shear Rate, Temperature, and Oil Concentration. Journal of Canadian Petroleum Technology 16:54-59.

Martin, S. 1979. A Field Study of Brine Drainage and Oil Entrainment in First Year Sea Ice. Jour. Glacial. 22:473-502.

Martin, S., P. Kauffman, and C. Parkinson. 1983. The Movement and Decay of Ice Edge Bands in the Winter Bering Sea. Jour. Geophys. Res. 88:2803-2812.

McAuliffe, C. D. 1977. Dispersal and Alteration of Oil Discharged on a Water Surface. Pages 19-35 In D. A. Wolfe, ed. Fate and Effects of Petroleum Hydrocarbons in Marine Organisms and Ecosystems. Pergamon Press, Inc., Elmsford, N.Y.

McAuliffe, C. D., D. E. Fitzgerald, B. L. Steelman, J. P. Ray, W. R. Leek, and C. D. Barker. 1981. The 1979 Southern California Dispersant Treated Research Oil Spills. Pages 269-282 In Proceedings, 1981 Oil Spill Conference, American Petroleum Institute. Washington, D.C.

McGowan, W. E., W. A. Saner, and G. L. Hufford. 1974. Tar Ball Sampling in the Western North Atlantic. Mar. Poll. Monitoring (Petroleum). National Bureau of Standards Spec. Pub. 409:83-84.

Mommessin, P. R. and J. C. Raia. 1975. Chemical and Physical Characterization of Tar Samples from the Marine Environment. Pages 155-167 In Proceedings, 1975 Conference on Prevention and Control of Oil Pollution. American Petroleum Institute. Washington, D.C.

Mooney, M. 1951. The Viscosity of a Concentrated Suspension of Spherical Particles. J. Colloid Sci. 10:162-170.

Morris, B. F. 1971. Petroleum: Tar Quantities Floating in the Northwestern Atlantic taken with a new Quantitative Neuston Net. Science, 173:430-432.

Morris, B. F. and J. N. Butler. 1973. Petroleum Residues in the Sargasso Sea and on Bermuda Beaches. Pages 521-529 In Proceedings, Joint Conference on the Prevention and Control of Oil Spills. American Petroleum Institute, Washington, D.C.

Morris, B. F., J. N. Butler, and A. Zsolany. 1975. Pelagic Tar in the Mediterranean Sea. Environ. Conserv., 2:275-281.

Morris, B. F., J. N. Butler, T. D. Sleeter, and J. Cadwallader. 1976. Transfer of Particulate Hydrocarbon Material from the Ocean Surface to the Water Column. In H. L. Windom and R. A. Duce, ed. Marine Pollutant Transfer. Published by D. C. Heath and Co., Lexington, MA.

Morris, R. J. and F. Culkin. 1974. Lipid Chemistry of Eastern Mediterranean Surface Layers. Nature 250:640-642.

Nagata, S. and G. Kondo. 1977. Photo-Oxidation of Crude Oils. Pages 617-620 In Proceedings, 1977 Oil Spill Conference American Petroleum Institute. Washington, D.C.

National Academy of Sciences. 1973. Petroleum in the Marine Environment. Workshop on Inputs, Fates, and the Effects of Petroleum in the Marine Environment (May 21-25, Airlie, VA) Washington, D.C. 824 pp.

Nelsen, T. A. 1980. Mineralogy of Suspended and Bottom Sediments in the Vicinity of the IXTOC-I Blowout, September 1979. Pages 189-204 In Proceedings of a Symposium, Preliminary Results from the September 1979 Researcher/Pierce IXTOC-I Cruise (June 9-10, Key Biscayne, FL) NOAA, Office of Marine Pollution Assessment. Washington, D.C.

Nelson, W. G. and A. A. Allen. 1981. Oil Migration and Modification Processes in Solid Sea Ice. Pages 191-198 In Proceedings, 1981 Oil Spill Conference. American Petroleum Institute, Washington, D.C.

Oil Spill Intelligence Report. 1980. An International Weekly Newsletter From the Center for Short-Lived Phenomena and Cahners Publishing Co., Boston, MA, Volume III, No. I, Jan. 4.

Overton, E. B., J. L. Laseter, S. W. Mascarella, C. Raschke, I. Nuiry, and J. W. Farrington. 1980a. Photo Chemical Oxidation of IXTOC-I Oil. Pages 341-383 In Proceedings of a Symposium, Preliminary Results from the September 1979 Researcher/Pierce IXTOC-I Cruise. (June 9-10, Key Biscayne, FL) NOAA, Office of Marine Pollution Assessment. Washington, D.C.

Overton, E. B., L. V. McCarthy, S. W. Mascarella, M. A. Maberry, S. R. Antoine, J. L. Laseter, and J. W. Farrington. 1980b. Detailed Chemical Analysis of IXTOC-I Crude Oil and Selected Environmental Samples from the Researcher and Pierce Cruises. Pages 439-495 In Proceedings of a Symposium, Preliminary Results from the September 1979 Researcher/Pierce IXTOC-I Cruise. (June 9-10, Kay Biscayne, FL) NOAA, Office of Marine Pollution Assessment. Washington, D.C.

Owens, E. H. 1978. Mechanical Dispersal of Oil Stranded in the Littoral Zone. Journal of the Fisheries Research Board of Canada 35:563-572.

Parker, C. A. 1970. The Ultimate Fate of Crude Oil at Sea; Uptake of Oil by Zooplankton. Page 242 In P. Hepple, ed. Water Pollution by Oil, Institute of Petroleum (London).

Parker, C. A., M. Freegarde, and C. G. Hatchard. 1971. The Effect of Some Chemical and Biological Factors on the Degradation of Crude Oil at Sea. Pages 237-244 In P. Hepple, ed. Water Pollution by Oil, Institute of Petroleum (London).

Payne, J. R., J. R. Clayton, Jr., B. W. de Lappe, P. L. Millikan, J. S. Parkin, R. K. Okazaki, E. F. Letterman, and R. W. Risebrough. 1978. Hydrocarbons in the Water Column, Southern California Baseline Study. Submitted to U.S. Department of Interior, BLM, as SAI Report No. SAI-76-809-LJ:149 pp.

Payne, J. R., G. S. Smith, P. J. Mankiewicz, R. F. Shokes, N. W. Flynn, V. Moreno, and J. Altamirano. 1980a. Horizontal and Vertical Transport of Dissolved and Particulate-Bound Higher-Molecular-Weight Hydrocarbons from the IXTOC-I Blowout. Pages 119-167 In Proceedings of a Symposium, Preliminary Results from the September

1979 Researcher/Pierce IXTOC-I Cruise. (June 9-10, Key Biscayne, FL) NOAA, Office of Marine Pollution Assessment. Washington, D.C.

Payne, J. R., N. W. Flynn, P. J. Mankiewicz, and G. S. Smith. 1980b. Surface Evaporation/Dissolution Partitioning of Lower-Molecular-Weight Aromatic Hydrocarbons in a Down-Plume Transect from the IXTOC-I Wellhead. Pages 239-265 In Proceedings of a Symposium, Preliminary Results from the September 1979 Researcher/Pierce IXTOC-I Cruise. (June 9-10, Key Biscayne, FL) NOAA, Office of Marine Pollution Assessment. Washington, D.C.

Payne, J. R., G. S. Smith, J. L. Lambach, and P. J. Mankiewicz. 1981a. Chemical Weathering of Petroleum Hydrocarbons in Sub-Arctic Sediments: Results of Chemical Analyses of Naturally Weathered Sediment Plots Spiked with Fresh and Artifically Weathered Cook Inlet Crude Oils. Final Report RU 190, Submitted to NOAA/OCSEAP, Juneau, Alaska. 367 pp.

Payne, J. R., B. E. Kirstein, R. E. Jordan, G. D. McNabb, Jr., J. L. Lambach, M. Frydrych, W. J. Paplawsky, G. S. Smith, P. J. Mankiewicz, R. T. Redding, D. M. Baxter, R. E. Spenger, R. F. Shokes, and D. J. Maiero. 1981b. Multivariate Analysis of Petroleum Weathering in the Marine Environment - Subarctic. Annual Report on Contract No. NABORAC00018 submitted to NOAA/OCSEAP Program Office, Juneau, Alaska. 364 pp.

Payne, J. R., B. E. Kirstein, G. D. McNabb, Jr., J. L. Lambach, C. deOliveira, R. E. Jordan, and W. Hom. 1983. Multivariate Analysis of Petroleum Hydrocarbon Weathering in the Subarctic Marine Environment. Pages 423-434 In Proceedings of the 1983 Oil Spill Conference. American Petroleum Institute, Washington, D.C.

Payne, J. R., B. E. Kirstein, G. D. McNabb, Jr., J. L. Lambach, R. T. Redding, R. E. Jordan, W. Hom, C. de Oliveira, G. S. Smith, D. M. Baxter, and R. Gaegel. 1984a. Multivariate Analysis of Petroleum Weathering in the Marine Environment - Subarctic. In: Environmental Assessment of the Alaskan Continental Shelf -- Final Reports of Principal Investigators. Vols. 21 and 22. U.S. Department of Commerce, National Oceanic and Atmospheric Administration, Juneau, Alaska.

Payne, J. R., G. D. McNabb, Jr., B. E. Kirstein, R. Redding, J. L. Lambach, C. R. Phillips, L. E. Hachmeister, and S. Martin. 1984b. Development of a Predictive Model for the Weathering of Oil in the Presence of Sea Ice. Final

Report prepared for NOAA, Outer Continental Shelf Environmental Assessment Program. Anchorage, AK. 253 pp.

Peak, E. E. and G. W. Hodgson. 1966. Alkanes in Aqueous Systems. I. Exploratory Investigations on the Accommodation of C-20 to C-33 n-alkanes in Distilled Water and Occurrence in Natural Water Systems. Journal of American Oil Chemists Society 43:215-222.

Peak, E. E. and G. W. Hodgson. 1967. Alkanes in Aqueous Systems. II. Combination of C-12 to C-36 n-alkanes in Distilled Water. Journal of American Oil Chemists Society 44:696-702.

Pequegnat, L. H. 1979. Pelagic Tar Concentrations in the Gulf of Mexico Over the South Texas Continental Shelf. Contributions in Mar. Sci. 22:31-39.

Percy, J. A. and P. G. Wells. 1985. Effects of Petroleum in Polar Marine Environments. Marine Technology Society Journal 18:51-61.

Petersen, H. K. 1978. Fate and Effect of Bunker C Oil Spilled by the USNS Potomac in Melville Bay - Greenland - 1977. Pages 332-343 In Proceedings, Conference on Assessment of Ecological Impacts of Oil Spills. (June 14-17, Keystone, CO) American Institute of Biological Sciences. Washington, D.C.

Polikarpov, G. G., N. Yegorov, V. N. Ivanov, A. V. Tokareva, and I. A. Feleppov. 1971. Oil Areas as an Ecological Niche. Priroda No. 11 (translated from Russian by Precoda, N.) Pollut. Abstract 3:72-5TC-0451. Cited in: Sleeter, T. D., B. F. Morris, and J. N. Butler, 1976. Pelagic Tar in the Caribbean and Equatorial Atlantic. 1974. Deep-Sea Res. 23:467-474.

Raj, P. P. K. 1977. Theoretical Study to Determine the Sea State Limit for the Survival of Oil Slicks on the Ocean. U.S. Department of Transportation, USCG Report No. CG-D-90-77.

Rashid, M. A. 1974. Degradation of Bunker C Oil Under Different Coastal Environments in Chedabucto Bay, Nova Scotia. Estuarine Coastal Marine Science 2:137-144.

Ross, S. L., C. W. Ross, F. Lepine, and R. K. Langtry. 1980. IXTOC-I Blowout. Pages 25-38 In Proceedings of a Symposium, Preliminary Results from the September 1979 Researcher/Pierce IXTOC-I Cruise. (June 9-10, Key

Biscayne, FL) NOAA, Office of Marine Pollution Assessment. Washington, D.C.

Sadd, J. L., E. R. Gundlach, W. Ernst, and G. I. Scott. 1980. Distribution, Size, and Oil Content of Tar Mats and the Extent of Buried Oil Along the South Texas Shoreline. Pages 14-20 In Research Planning Institute Report to NOAA, Office of Marine Pollution Assessment, Washington, D.C.

Saner, W. A. and M. Curtis. 1974. Tar Ball Loadings on Golden Beach, Florida. Mar. Poll. Monitoring (Petroleum). Nat. Bur. Stand. Spec. Pub. No. 409:79-81.

Shannon, L.V., P. Chapman, G. A. Eagle, and T. P. McClurg. 1983. A Comparative Study of Tar Ball Distribution and Movement in Two Boundary Current Regimes, Implications for Oiling of Beaches. Oil Petrochem. Poll. 1:243-259.

Shaw, D. G. and G. A. Mapes. 1979. Surface Circulation and the Distribution of Pelagic Tar and Plastic. Mar. Poll. Bull. 10:160-162.

Sherman, K., J. B. Colton, R. L. Dryfoos, and B. S. Kinnear. 1973. Oil and Plastics Contamination and Fish Larvae in Surface Waters of the Northwest Atlantic. MARMAP Operational Test Survey Report: July-August 1972, January-March 1973. Unpublished report. NMFS, MARMAP Narragansett, RI.

Sherman, K., J. B. Colton, R. L. Dryfoos, K. D. Knapp, and B. S. Kinnear. 1974. Distribution of Tar Balls and Nueston Sampling in the Gulf Stream System. Mar. Poll. Monitoring (Petroleum). Natl. Bur. of Stand. Spec. Pub. 409:83-84.

Sleeter, T. D. and J. N. Butler. 1982. Petroleum Hydrocarbons in Zooplankton Faecal Pellets from the Sargasso Sea. Mar. Poll. Bull. 13:54-56.

Sleeter, T. D., B. F. Morris, and J. N. Butler. 1974. Quantitative Sampling of Pelagic Tar in the North Atlantic. 1973. Deep-Sea Res. 21:773-775.

Sleeter, T. D., B. F. Morris, and J. N. Butler. 1976. Pelagic Tar in the Caribbean and Equatorial Atlantic. 1974. Deep-Sea Res. 23:467-474.

Smith, G. B. 1976. Pelagic Tar in the Norwegian Coastal Current. Mar. Poll. Bull. 7:70-72.

Smith, G. L. 1977. Determination of the Leeway of Oil Slicks. In D. A. Wolfe, ed. Fate and Effects of Petroleum Hydrocarbons in Marine Organisms and Ecosystems. Pergamon Press, Inc., Elmsford, NY.

Smith, S. R. and A. H. Knap. 1985. Significant Decrease in the Amount of Tar Stranding on Bermuda. Mar. Poll. Bull. 16:19-21.

Straughan, D. 1977. Biological Survey of Intertidal Areas in the Straits of Magellan in January, 1975, Five Months after the Metula Oil Spill. Pages 247-260 In D. A. Wolfe, ed. Fate and Effects of Petroleum Hydrocarbons in Marine Organisms and Ecosystems. Pergamon Press, Inc., Elmsford, NY.

Sweeney, R. E., R. I. Haddad, and I. R. Kaplan. 1980. Tracing the Dispersal of the IXTOC-I Oil Using C, H, S, and N Stable Isotope Ratios. Pages 89-115 In Proceedings of a Symposium, Preliminary Results from the September 1979 Researcher/Pierce IXTOC-I Cruise. (June 9-10, Key Biscayne, FL) NOAA, Office of Marine Pollution Assessment. Washington, D.C.

Thingstad, T. and B. Pengerud. 1983. The Formation of "Chocolate Mousse" from Statfjord Crude Oil and Seawater. Mar. Poll. Bull. 14:214-216.

Thomas, M. L. H. 1977. Long-Term Biological Effects of Bunker C in the Intertidal Zone. Pages 238-245 In D. A. Wolfe, ed. Fate and Effects of Petroleum Hydrocarbons in Marine Organisms and Ecosystems. Pergamon Press, Inc., Elmsford, NY.

Twardus, E. M. 1980. A Study to Evaluate the Combustibilty and Other Physical and Chemical Properties of Aged Oils and Emulsions. R and D Division Environmental Emergency Branch Environmental Impact Control Directorate Environmental Protection Service Environment, Canada, Ottawa, Ontario.

Vandermeulen, J. H., P. D. Keizer, and W. R. Penrose. 1977. Persistence of Non-Alkane Components of Bunker C Oil in Beach Sediments of Chedabucto Bay, and Lack of Their Metabolism by Mollusks. Pages 469-482 In Proceedings, 1977 Oil Spill Conference. American Petroleum Institute, Washington, D.C.

Vandermeulen, J. H., B. F. N. Long, and T. P. Ahern. 1981. Bioavailability of Stranded Amoco Cadiz Oil as a Function of Environmental Self-Cleaning: April 1978-January

1979. Pages 585-597 In Proceedings of the International Symposium, Amoco Cadiz Fates and Effects of the Oil Spill. Centre Oceanologique de Bretagne, Brest (France).

van Dolah, R. F., V. G. Burrell, Jr., and S. B. West. 1980. The Distribution of Pelagic Tars and Plastics in the South Atlantic Bight. Mar. Poll. Bull. 11:352-356.

Van Vleet, E. S., W. M. Sackett, F. F. Weber, Jr., and S. B. Reinhardt. 1981. Spatial and Temporal Variation of Pelagic Tar in the Eastern Gulf of Mexico. Adv. Organic Geochem. 4:362-368.

Van Vleet, E. S., W. M. Sackett, F. F. Weber, Jr., and S. B. Reinhardt. 1983. Input of Pelagic Tar into the Northwest Atlantic from the Gulf Loop Current: Chemical Characterization and its Relationship to Weathered IXTOC-I Oil. Canadian Jour. Fish. Aq. Sci. 40 (Suppl. 2):12-22.

Van Vleet, E. S., W. M. Sackett, S. B. Reinhardt, and M. E. Mangini. 1984. Distribution, Sources, and Fates of Floating Oil Residues in the Eastern Gulf of Mexico. Mar. Poll. Bull. 15:106-110.

Wade, T. L. 1974. Measurements of Hydrocarbon, Pthalic Acid, and Pthalic Acid Ester Concentrations in Environmental Samples from the North Atlantic, M.S. Thesis, University of Rhode Island.

Wade, T. L., J. G. Quinn, W. T. Lee, and C. W. Brown. 1976. Source and Distribution of Hydrocarbons in Surface Waters of the Sargasso Sea. Pages 271-286 In Proceedings of the Symposium, Sources, Effects and Sinks of Hydrocarbons in the Aquatic Environment. American Institute of Biological Sciences, Washington, D.C.

Walter, D. J. and J. R. Proni. 1980. Acoustic Observations of Subsurface Scattering During a Cruise at the IXTOC-I Blowout in the Bay of Campeche, Gulf of Mexico. Pages 525-541 In Proceedings of a Symposium, Preliminary Results from the September 1979 Researcher/Pierce IXTOC-I Cruise. (June 9-10, Key Biscayne, FL) NOAA, Office of Marine Pollution Assessment. Washington, D.C.

Ward, D. M. 1981. Microbial Responses to Amoco Cadiz Oil Pollutants. Pages 217-222 In Proceedings of the International Symposium, Amoco Cadiz Fates and Effects of the Oil Spill. Centre Oceanologique de Bretagne, Brest (France).

Weeks, W. F. and G. Weller. 1984. Offshore Oil in the Alaskan Arctic. Science 225:371-378.

Wolfe, D. A., R. C. Clark, Jr., C. A. Foster, J. W. Hawkes, and W. D. MacLeod, Jr. 1981. Hydrocarbon Accumulation and Histopathology in Bivalve Molluscs Transplanted to the Baie de Morlaix and the Rade de Brest. Pages 599-616 In Proceedings of the International Symposium, Amoco Cadiz Fates and Effects of the Oil Spill. Centre Oceanologique de Bretagne, Brest (France).

Wong, C. S., D. R. Green, and W. J. Cretney. 1973. Pelagic Tar in the North Pacific Ocean. Pages 400-415 In Background Papers for a Workshop on Inputs, Fates, and Effects of Petroleum in the Marine Environment, Volume 2. Prepared under the Aegis of Ocean Affairs Board, National Academy of Sciences, Washington, D.C.

Wong, C. S., D. R. Green, and W. J. Cretney. 1974. Quantitative Tar and Plastic Waste Distributions in the Pacific Ocean. Nature 247:30-32.

Wong, C. S., D. R. Green, and W. J. Cretney. 1976. Distribution and Source of Tar in the Pacific Ocean. Mar. Poll. Bull. 7:102-106.

Zajic, J. E., B. Supplisson, and B. Volesky. 1974. Bacterial Degradation and Emulsification of No. 6 Fuel Oil. Environ. Sci. Technol. 8:664-668.

INDEX

Aber Benoit 56,57
Aber Wrac'h 52,54,55
acoustic profiling 65
acid content 11,21,97
activity coefficients 93
advection 94
Agah Jari (see Iranian Light Crude)
agglomeration (see mousse agglomeration)
aggregation 3,69,99
air/oil interface 110
air/sea interface 94
Alberta Crude 7,15
Alboran Sea 85
alcohols 23
aldehydes 23
Algeria Crude (See Zarzatine)
algorithms 80,99–111
alkyl peroxide radicals 10,23
alkyl radicals 23
Alvenus 42,77–78
amino acid uptake 74
Amoco Cadiz
  oil 53,57
  spill 39,42,51–57,117–119
Anchorage, Alaska 117
anoxic conditions 53
anthracene 14,52
  rings 14
anthropogenic pollution 49
Antifer (Le Havre) 38
Antilles 84
aqueous phase 38
Arabian Light Crude 6,13,14,23,37,38,46, 49,113,117
Arabian Sea 85,91
Argo Merchant 65
aromatic compounds 14,16,32,45
arragonite 61
Arrow 42,43–46,56,91,93,117,118
artificial light 24
asphaltenes 4,6–8,11,12,21–24,32,42, 45,47,56,86–88,97,113,114,117
Atlantic Ocean 84,86,87,89,118
  Central 84
  North 83
  Northwest 83,87
  South Atlantic Bight 84,89
Auk Crude 9
auto-oxidation 22,23,38

bacteria 11,21,40,70,74,89,90,114,115
bacterial growth 6–8,115
bacterial products 49
bacterial slime 21
bacterial utilization 40,89,92,115
Baffin Bay, Greenland 50
Bahamas 84
Balearic Sea 85
Barbados 95
Barents Sea 84
barnacles 87,92
  larvae 43
Basra Crude 37
Bay of Campeche 42,57–76,91,117
Bay of Morlaix, France 55
beaches 82,92
  fore-dune ridge 75
  high beach zone 75
  sandbar 75,92
  shoreline break 75,92
  swash zone 75,92,96
Beaufort Sea 95
Bellvue 103
benthic fauna 55
benzene 26–29,66,78,89
  alkyl benzenes 50
  alylic benzoic acids 69
  tetramethylbenzene 52
benzoanthracene 14
  benzo(a)anthracene 2
benzofluoranthrenes 55
benzopyrenes 55
  benzo(a)pyrene 14
benzothiophenes 14,24,70
  alkyl benzothiophene acids 69
Bermuda 84,95,96,97
beta-carotene 23
biodegradation 53,54,87,88
bi-wetted solid 35
Blokker equations 108

blue-green algae 91
Boehlen Spill 40
Bolinas Bay, California 92
Bolivar peninsula 47
Boston 78
Bow River Crude 6,15
BP 1002 33,36
Brazos Island, Texas 74
Breaxit 33,36
Brega (See Libyan Crude)
Brent Crude 9,10,36,37
brine channels 20,32
Brownian Movement 4
Brownsville, Texas 57–59
bryozoans 87,91
bubbles 10,83
Bunker C Crude 6,10,14,32,36,37,42,43,
45,46,51,81,91,92,109,110
buoyancy 87,109
Burmah Agate 42,76–77,117
burnpoint 30,115
butane 25–27
methyl 26–27

Cabimas Crude 9
Cabo Rojo (Tuxpan de Rodriguez Cano)
57
Calanus finmarchicus 44
calcite 60
California Crude 4
Cameron, Louisiana 77
Canadian Environmental Protection Service
58
Candida petrophiles 22
C. tropicalis 22
capping 57
carbonate minerals 60,87
carbon compounds 84
carboxylic acids 23
Caribbean Sea 85,95
case histories 41–82,116–118
Cerberus Rock (See Chedabucto Bay)
Chedabucto Bay, Nova Scotia 43,46,56
chitin 94
chlorite 60
chlorosis 43
chromatogram profiles 26–27,43,45,52,
53,82,91
chronic petroleum pollution 48
chrysene 14,52
Chuckchi Sea 19
Ciudad del Carmen 57,59
clean-up operations 20,33,43,59,69,116
climatic differences 90
coalescence 3,15,21,100,102,104,105,
113,114
coastal embayments 53
Coatzacoalcos, Mexico 57
colloidal particles 42,91
combustibility 30–32,115
combustion 20,115
compressive forces 109
computer plots 29,31
computer simulations 32,51,99–111
contamination potential 92
continuous phase (external phase) 3
controlled spills 40,78–81
convergence zones 69
Cook Inlet, Alaska 40
Cook Inlet Crude 94
copepods 43,44,93
gut contents 42
Corexit 8666 43
Corexit 9527 28
Cornwall coast 41
Corpus Christi 58
Crichton Island, Nova Scotia 45
cumene 23
currents 59,61,95
coastal currents 92
eddy currents 62
estuarine 92
subsurface currents 92
cyclohexane 26–29
methyl 26,27
cyclopentane 26,27

degradation rates 94
density 30,78,80,109
changes 15,16
gradients 65
detergents 35
detritus 69,92
diatoms 91
dibenzothiophenes 14,24,48,52,54,55,70,
71
alkyl-dibenzothiophene 52
dibenzothiophene acids 69
dichlorophen 10
diesel oil 10,15,114
auto 8
marine 8
differential equations 102
diffusion 105
coefficients 89
control 93
rates 88,89,104
diffusivities 89,104,111
vertical diffusivity 104
dilution 81
discharges (see waste products)

dispersants/demulsifiers 33–39,43,79,80, 104,115,116
  aerial application 76,77
  spray boat application 77
dispersed (internal) phase 3
dispersion 19,25,33,82,86,99,104,115, 120
  model 104,107,109
  rate 46,107
dissolution 12,18,25,33,40,50,52,53,60, 66,69,79,82,87,99,115,116,118,120
  rate 109
dolomite 60
drag
  coefficient 109
  profile 81
Durelle Island 45
dye-marker studies 80
dysprosium lamp 23

Ekofisk Bravo
  blowout 42,47–50,65,80,91,94,117
  oil 9,47,101,102
electro-kinetic potential 4
emulsification energy 3,93
emulsifying agents 4,21–24,99
emulsion breakers 39
energy regimes 82,118,119
environmental impacts 48,80,104,117
Escaros 103
Esso Breaxit (see Breaxit)
ester content 97
estuaries 46,82,118
estuarine marsh systems 53
evaporation 12,13,15,18,25–30,38,40, 49–53,60,66,67,69,78,79,82,93,96, 99,108,109,110,117,118,120
evaporation-in-pan experiments 14,15
evaporative loss 47,65,66,78,92
  weathering model 92,93

Falmouth Harbor 96
fatty acids 69
fatty acid methyl ester 69
fecal pellets 43,93,117
feldspar 61
ferric oxide (rust) 90
field tests (see controlled spills)
fire 61
  point 30
flash point 30,115
floating debris 83
Florida coast 85,86,88,92,95
flotsam (beached debris) 76
flow properties 9,100
fluoranthenes 55,71
fluorene 14,52,70
fluorescence spectra 43
fluorescent lamp 23
fluxes 104
formation rates 102
fragmatograms, GC/MS 47
free radical chain 23
freeze/thaw cycles 33,114
fuel oils 15,37
  #2 fuel oil 80,81
  #4 fuel oil 80,81
  #6 fuel oil 22,80,81

Gach Seran (See Iranian Heavy Crude)
Galveston, Texas 76,77,87
Galveston Island 77
gas chromatography 16,20,45,52,54,55, 86,89
gasoline 7,10,64,114
geomorphology 53
glass capillary flame ionization detector gas chromotography 25–26
Greenland 50,81,109
Guanipa Crude 7,12
Gulf of Mexico 24,49,57,60,70,76,86,89, 117
Gulf Stream 85

heavy fuel oil 14
heavy oil A 13
helicopter overflights 67
Henry's Law 111
hetero-aromatics 24
hexadecane 22
hexane 26–29
H-nme 86
hopanes 55
  diastereomer 49
Hurricane Allen 75
  Henri 60,64,65,67
hydroxyl content 97

ice 18,20,30,32,104,114
  columnar 19
  first year 19
  frazil 19
  grease 19,21
ice break-up 19
ice floe 19,20,114
ice growth 19
ice thaw 19
Idothea metallica 92
ignition 20,30
ignition times 30
Ile Grande Salt Marsh 5
illite 61

incubation period 72
India coast 95
industrial fuel oils 18
infrared (IR)
  absorption 86,88
  measurements 38,86
  spectrometry 84,88,89
ingestion 88,117
Inhabitants Bay, Nova Scotia 45
inorganic salts 89
interfacial area 3,99
  film 3
  surface tension 3,16,17,19,99,107,114
intertidal energy 54
  substrate 42,43,45,118,119
  zone 45,47,70,74,77,82,97,118,119
Ionian Sea 85
ions
  mono-valent 4
  poly-valent 4
Iranian Crudes
  light (Agah Jari) 6,10,11,13,36
  heavy (Gach Seran) 6,9,11,35,36
Iraq Crude (Kirkuk) 6,11,36
Irene's Serenad Spill 37
iron 90
isopods 87
isoprenoids 23,52,54
Israel coast 95
IXTOC-1
  crude 60,65,67,69–72,74,76,81
  mousse 59–61,74
    plume movement 59
  reservoir 60
  spill 12,13,24,42,57–76,79,82,87,91,92,94,117–119
  well 67,70,71,72,75,118

JP-5(aircraft fuel) 115

Kaolinite 61
Kasitsna Bay, Alaska 18,26–28,31,33,37,89,94
kerosene 7,10,114
ketones 23
kinematic viscosity 5,8,11
Kirkuk (See Iraq Crude)
Kovat indices 29,30,31
Kuwait Crude 6,7,9–11,13,21,22,35-37,40,41,46,114,116
  mousse 11,35

LA 1834 36,37
laboratory studies 3–40,78,113–115
lagoons 45,59,82,118
laguna de Terminas 59,62
Laguna Madre 59,69,70,72,73,75,118
landfalls 78
Langmuir circulation 63,64,104
La Rosa Crude 78–80,116
lead matrix pumping 19
leeway 80
lenses 78,79,116
Lepas pectinata 92
Libyan Crude 5–8,10,11,22,35,103,113
Lloydminster Crude 7,15,32,33
Long Beach, California 79
long-term contamination 44
Loop Current 85
lube oils 7,10,22,39,114
  lube oil 600 8
  lube oil 2500 8
  heavy naph lube 8
  refined 21

magnesium calcite 61
manufacturing processes 84
marsh detritus 64
Martha's Vineyard 97
mass balance 12,32,51,93
mass transfer coefficient 105
Matagorda Bay, Texas 58
mediterranean Sea 83,85,86,118
melting (thawing) behavior 18
Melville Bay, Greenland 50
Merey crude 42,77,78
metal content 5,10
metalloporphyrins 21,107,113
methylethylketone/dichloromethane 22
methylpentane 26,27
Metula Spill 42,46–47,91,116,118
Mexican coastline 55,57,58
microbes (See bacteria)
microbial degradation 12,13,32,33,38,50,54,70,89,93,94,97,104,118,119
microbial oxidation 90,113
microbial processes 52
microcosm experiments 70,74
microparticulates 38,41
microscopic examination 14,49
Middle East Crude Oil 22
Mimosa (freighter collision) 76
mineral diffusion 12
mineral species 61
mixing chambers 15
mixing energy 23,29,60,99,116
mixing regimes 81
MLB 103
models/modeling 15,18,80,93,99–111,119–120
  evaporative weathering 92
  mathematical 97

oil spill 101
oil weathering behavior 78,119
trajectory 79
verification 105
water-in-oil emulsion 15,101
monitoring 90
montmorillinite 61
Mooney equations 100
mopping 115
mousse agglomeration 67,68,74,119
artificially generated 33
dispersion 39,93
mousse
flakes 70
formation 3–40,60
physical properties 115
resolved components 34
stability 18,21,113
structure 113
unresolved complex mixture 34
viscosity 95,113
mud flats 46,53,54,56
Murban Crude 78–80
mussels 52

n-alkane/isoprenoid ratio 55
naphthalenes 48,50,52,70,71
alkyl-substituted 66
C_3 naphthalene 52
dimethylnaphthalene 52
methylnaphthalene 52,78
naphthanoic acids 69
naphthenes 32
naphthols 69
Narragansett Bay 91
National Oceanic and Atmospheric Administration (NOAA) 24,59,60
oil spill trajectory models 57
nC-17/pristane ratio 50,52,74,82,83,87,89
nC-18/phytane ratio 50,52,70,73,74,87,89
nC-18/nC-19 ratio 49
nC-25/nC-16 ratio 70,73
nC-25/nC-19 ratio 70,73
nC-27/nC-26 ratio 49
neuston 119
New Jersey 79
Newtonian behavior 100
New York Bight 88
Harbor 83,87,88
nickel 6–8,11–13,16
Nigerian Crude 35,36,76,113,117
light 5,6,10,11,14,22,31,77
medium 9
nitrogen 53,56,92
nitrogen, sulfur, and oxygen (NSO) compounds 56,87,89

nonane 26,27
Norman Wells Crude 6,15,32,33
North Brittany, France 42,51,118
North Sea 46
NPD (See dibenzothiophene)
nutrients 11,40,74,94,115,119
concentrations 89
depletion 40,74
diffusion 115,119
utilization 74

oceanic fronts 69
oceanographic conditions 48,81
octane 26,27
oil
density 106,108
droplets 3,43,50,61,66,67,69,104
flakes 46,66,67
pancakes 50
pellets 22
precipitates 86
resolved components 34
slicks 59
sludge 89,90
unresolved components 34
oil spill transport 76
oil tankers 83
bilge pumping 90,91,119
discharges 90,96
load-on-top procedures 90
traffic routes 118
washing and ballasting operations 88, 89,90,119
Oil Vulnerability Index 53
oil/water interface 24,78,110,114
Oman coast 95
oxidation 14,87
biological 87
chemical 87
products 24
oxygen 23,53,56,88,92,113,119
compounds 38
consumption 40
content 11
depletion 11,40,53
diffusion 12
oxygenated products 12,23

Pacific Ocean 83,85,87
eastern 83,85,118
north 83
south 85
western 83
Padre Island, Texas 59,74
pancreatic lipase 22

paraffins 32
  n-paraffins 13
  secondary 14
  tertiary 14
particulate materials (See suspended particulate material)
particulate oil/particulate mousse 52
pelagic organisms 87
pentacyclic triterpanes 54
pentane 22,26–29
peroxides 23
perylene 89
Petroleos Mexicanos (PEMEX) 57,60
petroleum
  sludge 89,90
  storage 90
  transport 90
petroleum to biogenic hydrocarbon ratio 94
phase separation 35
phenanthrene 14,48,52,70,71
  alkyl-substituted 53
  methylphenanthrene 50,52
phenanthroic acids 69
phenols
  alkyl-phenols 69
phosphate 11
phosphorus 53
photochemical decomposition (degradation) 14,22,23,33,38,60,69, 119
photo-micrographs 35
photo-oxidation 23,49,69,90,113,116
phytane 54,74,89
GW Pierce Cruise 60–62,64,67,68,71, 72,74
Pilon crude 42,77,78
pipelines 49,90
pitch 111
polar compounds 12
polar groups 23
polymerization 23,24
polynuclear aromatic hydrocarbons 45,55, 70,71,89
  alkyl-substituted 69
polypeptides 22
polysaccharide 22
porphyrin complexes 4
Port Aransas, Texas 59
Port Mansfield 59
Portsall, France 55
Portsmouth, England 11
Potomac US/UN Spill 13,42,50–51,81, 91,109,110,117,118
pour point 5–8,11,12,30,115
predictive capabilities 102
pristane 54,74,89
pristane to phytane ratio 52
Product "A" 38
propane 26,27
protein 22
proton extraction 23
Prudhoe Bay, Alaska
Prudhoe Bay Crude 7,15–17,25,26–28, 30–33,79–81
Pseudomonad 22
  Pseudomonad aeruginosa 22
Puerto Espora, Mexico 46
pumping efficiency 115,116
pyrelene 55
pyrene 14,18,55,70,71

Qatar Marine Crude 9
quartz 61,87

Rabbit Island, Nova Scotia 45
rain 14,61,104
rate of propagation 20
relief well 57
Research Planning Institute 58
Researcher Cruise 24,58,60–64,72,74
residence time 43,88
resin fractions 88
resonance stabilized 23
rheological properties 16,17,19,100
rising velocity 104,105
river flow 59
rocky shorelines 54

Safanya Crude 14,37
salinity 10
saltwater/freshwater interface 59
sand 46,76,96
  erosion 76
  flats 54
San Francisco Bay, California 92
Sansinena Spill 32
Sargasso Sea 83,84,86,87,91,93,96,118
Sargassum 87
Satellite Patch (Straits of Magellan) 46
satellite photographs 56
saturate to aromatic hydrocarbon ratio 83, 97
scavenging 63,65
sea state 100,101,106
seaweed/algae 46,74,97
sediments 45,49,60,67,75,76,80,97
  disturbance 48
  extracts 54
  intertidal 45
  resuspension 60

sedimentation
processes 67,82
rates 93,110,119
regime 94
seep oils 82,87,90
selected ion monitoring GC/MS 66
sensitive biological habitats 59
settling time 34
shear rate 4,14,100
sheen 50,57,59,64,69,75,78,93,108–111,116
formation 51
shrimp nurseries 59
sinking 12,92
sites of nucleation 65,69,90
skimmers 39,115
slicks 60,61
diameter 108
life times 106
stretching 104
thickness 32,51,100,106–108
smectite 61
Smith-MacIntyre grab 49,67,94
snowfall 14
sodium chloride 20
sorbants 33,115
hydrophobic properties 115
sorption capacity 33,115
sour blend crude 6
Southern California 76,95
Spartina alterniflora (cord grass) 44
specific gravity 5,8,11–12,25,32,43,90,115
spill response countermeasures 98
SPM-bound oil 64
spreading 25,46,79,99,107,115
experiments 108
force 109
properties 108,115
stabilizing agent 94,113,116
standing waves 69
starch 94
static equilibrium 109
stirring time 37
St. Michael-En-Grave, France 55
Stokes Law 110
storage tanks 39
storm activity 60,61,64,75,76,92,94,96
Stratford crude 23
Straits of Magellan 42,46
stranded mousse (beached mousse) 11,37,39,55,72–74,118
stranded tar balls 94–97
concentration 95
distribution 95
persistence 119
subarctic environment 94
weather 15,94
subarctic weathering 16,25
substrate type 54
subsurface blowouts 65
subsurface oil 60
plume 67
release 66
transport 65
subtidal substrate 44
sulfoxides 23,24
formation 23,24
sulfur compounds 23,54,88,113
content 5–8,10,11,12,21,88
sunlight intensity 63
surface active compounds 18,23,35,81,113,114
surface area 97,99,109
surface flares 61
surface flocculate material 49
surface tension 17,99,109
surface to volume ratios 24,67,92,115
surfactants 3,4,18,20,28,35,37,96,113
suspended particulate material (SPM) 25,60,61,65,67,92,114
Sweet Blend Crude 6,15,32,33
synchronous excitation emission fluorescence scanning spectroscopy (SEES) 45
synthetic emulsions 38

Tampico, Mexico 57,58,59
tanks/tank systems 12,13,25,39
flow-through seawater 11,18,33
tests 38
wave generator 12
Tanyo Oil Spill 40
tar balls 12,13,46,75,81–97,118–119
chemical composition 86–90
dating 93
decomposition 119
dispersion 86
fate at sea 91–94,119
formation and distribution 83–94,118–119
standing crops 79,82,119
tar ball pollution 119
tar mats 75,77
temperature 4,114
Ten section 103
terrigenous runoff material 55
tertiary C-H bonds 14
n-tetracosane 50
tetradecanal 23
tetralin 23
Texas A&M University (TAMU) 58

Texas
beaches 86,92
coastline 59,60,74–77,87
thiacyclanes 24
thiacyclane oxides 24
Tia Juana (Venezuelan) Crude 5,6,10,11, 12,21,36,77–79
tide pools 56
tidal flats (See mud flats)
tidal
action 51,75
currents 77
Tierra del Fuego 46
Ti Saozin Island 56
toluene 26–29,66
Torrey Canyon Spill 24,41–43,116,118
total organic carbon (TOC) measurements 38
toxic effects 43,44,70
toxic materials 11,74,115
transfer coefficients 110,111
transfer rates 110
turbid/clear-water boundary 61,62
turbulence 25–29,46,52,77,91,104,116
Tuxpan de Rodriguez 59
Tyrrhenian Sea 85

ultraviolet (UV) light 23
unresolved complex mixture (UCM) 24, 34,45,54,86,89,97
U.S. Coast Guard Stations
Echo/Delta 87
November 87
USCG Westwind 50

valence 23
vanadium 5–8,11–13,16,21,23
vapor pressure 92,109
Venezuelan Crude (See Tia Juana Crude)
Veracruz, Mexico 57,58,59
Vertical flux 93
Vibrio spp. 70
viscosity 4,6–8,12,13,17,25,37,39,49,60, 77,78,81,90,99,100,102,103,107,109, 111,114–116,120
viscosity, effects on dispersants 39
visible light 20
volatile hydrocarbons 15,33

Wabern-Midale Crude 7,15,32,33
waste disposal trucks 37,39
waste products (discharges)
industrial wastes 88,119
urban wastes 88,119
water column 43,60,70
water content 6–8,12,13,17,37–39,78, 99–103,114,115,120
water droplets 18,20,31,81,114,115
water temperature 51,109
water uptake/incorporation 15,47,81,97
waves/wave action 3,39,51,75,77,96,116
breaking (white capping) 104
dispersion 106
energy 13,44,69
height 51,109
wave generator 12
wave tanks 5,15,16,33,37
studies 81
waxes 4,20,88,113,114
chrystallizing properties 113
wax content 6–8,10,11,19–22
wax inclusions 90
wetting agent 35
wind 3,48,51,60,61,63,75,80,83,96
direction 74,77
speed 101,102,107,109
wind rows 50,69,77

xylenes 26,27,30,78
meta 66
ortho 66
para 28,29,66

yeasts 22
Yucatan Peninsula 59

Zarzatine Crude 7,14,23,113
zooplankton 43,93
feces 43,93
Zveitna 103